TRISMS

Discovering the Ancient World

Creation - 500 B.C.

User's Manual

Sally Barnard

Linda Thornhill

Self-published by
TRISMS
1203 S. Delaware Place
Tulsa, OK 74104-4129

Other titles by TRISMS, Inc.
The History Makers
Rise of Nations
It's About Time
Reading Through the Ages

Current listing and prices are available from TRISMS, Inc.

Phone: 918-585-2778
Web site: www.trisms.com
Email: Linda@trisms.com

Self-published by TRISMS, Inc.
1203 S. Delaware Place
Tulsa, OK 74104-4129

Cover design by Kallen Godsey

Printed in the USA

Written for the inquiring minds of
Micah, Jared, Luke, and Sarah Barnard
Aaron, Jesse, and Daniel Thornhill

Many thanks to Cheryl Horton for her many hours
of typing, editing, and encouragement.

TABLE of CONTENTS

Time Tables: This table lists the civilizations and time periods covered in TRISMS *Discovering the Ancient World*.

Orientation: This gives an introduction and instructions for TRISMS. It defines how students and teachers will use the curriculum.

Resource & Reading List: This section contains reading selections. These are divided into resource books, junior level resource, historical fiction, periodicals, videos, and audiotapes. We realize not all libraries carry the same books, etc. This is a topical study so use the list as a guide for selecting similar resources.

Unit Plans: This is the user's guide for the school year. It is set up in a unit-at-a-glance format with assignments.

Science: science assignments and activities

Literature Assignments: assignments for literature and composition

Literature Selections: The pictures, poetry, drama, essays, short stories, and oratories from early civilizations to be used with specific assignments.

Literature Helps: This section includes grading tips and definitions of literary terms that will be helpful for completing Literature assignments.

All Questionnaires, Worksheets, Maps, Unit Quizzes, Vocabulary Quizzes, and Semester Tests are packaged together in the Student Pack.

Answer keys are provided for all Worksheets, Science assignments, Questionnaires, Literature, Unit Quizzes, Vocabulary Quizzes, and Semester Tests in the Answer Keys book.

TRISMS©

Time Tables

TIME TABLES

Ancient

Unit 1   Orientation
Creation/ Deluge/ Dispensation

Unit 2   Sumerians c. 4000 - 1750 B.C.
Indus Valley c. 3000 - 1500 B.C. (short form)

Unit 3   Ancient Egypt c. 4000 - 2686 B.C.
Old Kingdom c. 2686 - 2155 B.C.

Unit 4   Middle Kingdom c. 2040 - 1786 B.C.
New Kingdom c. 1550 - 1085 B.C.

Unit 5   Phoenicians c. 3000 - 860 B.C.
Carthage 850 - 146 B.C. (short form)

Unit 6   Minoan c. 3500 - 1400 B.C.

Unit 7   China
Shang c. 1500 - 1027 B.C. (short form)
Chou c. 1027 - 257 B.C. (short form)

Unit 8   Hittites c. 2000 - 1150 B.C. (short form)
Aryan c. 2000 - 600 B.C.

Unit 9   Old Babylon c. 2000 - 1600 B.C.
Special Project -- Mysteries of the World
Semester test

Unit 10   Mycenaean c. 1600 - 1120 B.C.

Unit 11   Olmec c. 1200 - 500 B.C. (short form)
Nubia c. 2000 - 200 B.C. (short form)

Unit 12   Hebrews -- Nomadic period -- Exodus to 1025 B.C.

Unit 13   Hebrews -- The Kingdom c. 1025 - 926

Unit 14   Ancient Greece c. 1150 - 500 B.C.
Dark Ages c. 1150 - 750 (short form)
Archaic c. 650 - 500 B.C. (short form)

Unit 15   Assyria c. 1500 - 612 B.C.

Unit 16   Etruscan c. 800 - 400 B.C.

Unit 17   Neo-Babylonian c. 612 - 539 B.C.

Unit 18   Scythians 800 B.C. - A.D. 200 (short form)
Special Project -- Wonders of the Ancient World
Semester test

Orientation

This section will introduce and explain the components of the TRISMS curriculum. Here you will learn how to use the lesson plan, make the timeline, build the coursebook, utilize blank maps and tie everything together. You will also learn what additional resources you will need to help you make full use of this course of study.

OBJECTIVES and OVERVIEW

The TRISMS curriculum is designed to impart a chronological survey and geographical progression of man over time. It is our goal to equip the students with skills that will prepare them for a lifetime of learning. TRISMS is more than a history study. It is designed to build upon a foundation of research, discovery, and language. One of the primary goals is to teach students to ask questions, find answers, and transfer information from reading to thinking, to writing, and to speaking. In order to stimulate the student's interest; TRISMS *Discovering the Ancient World* emphasizes humanities. All music, art, architecture, literature, and science come from the civilization being studied. This integration and immersion work together to help the student understand the people studied as well as make the information being learned more meaningful. The immersion of the student is doubly important for breaking through the 21st century "instant everything" mentality. Throughout the year the student will read literature, biographies, historical fiction, cultural studies, and most varieties of reference materials. Some students find the subject matter especially interesting and will want to spend more time than the unit plan presents. In addition, TRISMS is easily adapted to a multi-level setting. These features will be presented in more detail later in the orientation.

At the heart of the curriculum is the unit plan. The unit plan is organized in an overview format with assignments. Art, Music, Architecture, Science, Civilizations, Literature, and Vocabulary are covered in chronological order from the Beginning up to 500 BC. The unit plan also provides items of historical significance not covered by daily assignments; it suggests additional subjects of interest for more motivated students, and offers a comparison question.

As the student works through the unit plan, they will construct a timeline that chronologically lists the civilizations being studied by geographic area. The timeline provides a big picture overview and reference that relates to the subject matter, thus cultivating a broad understanding of the time period being studied. In addition to creating a timeline, the student will create a coursebook for Civilizations. This will contain the questionnaires that the student has completed and provide an organized reference for the subjects studied. Also included will be maps, drawings, articles, and pictures the student gathers during the course of study. Blank world maps will be used as a visual example of civilizations' growth and expansion.

The following sections explain how each part of *TRISMS* is used.

UNIT PLAN

Unit Plans are set up in a unit-at-a-glance format with activities in categories for Art, Music, Architecture, Science, Civilizations, Literature, Maps details, and Vocabulary. Historical Events lists items of historical significance in order to broaden the understanding of the period being studied. Map Details are to be labeled on the map(s) for the unit. Other Areas of Interest gives ideas for the student interested in learning more. A Compare Question is given to help develop critical thinking and abstract reasoning. Each unit is designed to take two weeks or more, depending on interest.

Note: Mathematics is not included in the *TRISMS* curriculum. The authors recommend Saxon Math as a supplemental text to provide for this need.

UNIT ASSIGNMENTS

Assignments are given in the unit plan for Art, Music, Architecture, Science, Civilizations, Worksheets, Literature, Maps, and Vocabulary. Questionnaire forms are provided for Art, Music, Architecture, and Civilizations. The student will use reference materials such as encyclopedias, books, periodicals, videos, and maps to fill out the questionnaires. The vocabulary words listed are words the student will come across in their reading and study of the subject. They should record the word, pronunciation, definition, and use the word in a sample sentence. The student will keep these words with each civilization studied and will be quizzed by unit on them. Selected words will be on the semester test.

A study guide can be made by using coiled index-cards noting words on one side and definitions on the opposite side.

Vocabulary words are also included with Literature assignments but will not appear on tests.

Additional assignments in Science and Literature are referenced in the unit plan and explained in the Science Section and Literature & Composition Section of the User's Manual. *The Institute for Excellence in Writing* program is used to develop structure and style. It is not required to do the assignments, but highly recommended. It is available through TRISMS.

Civilizations

Civilizations will be studied in a survey format using the Civilization Questionnaire or Short Form. A map is provided for each civilization studied. The extent of the civilization or empire should be marked on the map. These maps are blank so students must label the geographic region of the civilization. The map should include: mountain ranges, peninsulas, deserts and bodies of water: seas, oceans, lakes, straits, bays, and rivers. Also mark the ancient capitol, major cities, and ports. The maps are numbered and a notation is given with each civilization identifying which map to use. Each civilization will have an assigned worksheet to bring out information unique to these people. The worksheets are noted in the unit plan under Civilizations, and are located in the Student Pack. Answer keys are found in the Answer Keys book.

Worksheets

Worksheets look for information specific to the civilization being studied. While reading historical accounts, it is important to keep in mind that the victor records history and therefore it is biased. Reference books also tend to view history through hindsight or from a 20th or 21st century view, summing up the cause and effect of the situation. You must evaluate the situation as if you are there.

UNIT ACTIVITIES

Compare Question

The Compare Question is given to develop critical thinking and abstract reasoning. It is intended to help give the student a broader understanding of civilizations' relationship with each other by comparing how they are similar and different. These questions can be used for discussion, debate, or an additional essay assignment.

World Map

A blank world map is used to reinforce the idea of going back to the beginning of time, no countries or boundaries. The student should note weekly the geographic progress of civilization. They should become aware of the significance of rivers and mountains on expansion, as well as the terrain's effect on the development of transportation. Mark all coexisting civilizations on the notebook size world map that is assigned in the unit.

Movies

Movies can be used to further immerse the student in the time period. Most films illustrate costume, custom, social structure, architecture and often language forms. We suggest viewing movies prior to the study of the civilization or time period to promote anticipation. Films are listed in the Resource and Reading Selections section with a V for video.

Oral Reading

This optional reading gives an opportunity to read for pleasure selections that are not chosen for study.

Oral reading can be done by the teacher or in a round robin fashion with everyone taking a turn. Biographies and historical fiction are listed in the Resource and Reading Selections section or they can be exchanged for others from the time period.

TIMELINE

A timeline is a long strip of paper on which a series of dates are printed at specific intervals. It is a visual representation of consecutive events, a graph. The student will record all civilizations studied on their timeline. They are encouraged to add persons or nations they read about in addition to the assigned civilizations. The student may note that different people groups have their own calendars. Some are based on the solar year and others on the lunar year.

BC & AD

In using a timeline, you must know how to read dates and understand their meaning. To better comprehend history, we need to arrange events in order. Time is counted backward and forward from the day Christ was born. The years before that date are numbered backward, and each year is followed by the letters BC, meaning "before Christ." The years after Christ's birth are numbered forward and are preceded by the letters AD, which stands for "anno Domini". This is translated to mean "in the year of our Lord."

An event that took place in 10 BC happened ten years before Christ was born. Something that occurred in the year AD 100 took place 100 years after Christ's birth. Usually only the letters BC are used. Dates without letters are known to be "anno Domini".

A century is one hundred years. Therefore, the date 25 BC is in the first one hundred years BC, or the first century before Christ was born. An event in 170 BC happened in the second one hundred years, or the second century. An event in AD 350 was in the fourth one hundred years, or the fourth century.

Constructing the Timeline

Fanfold computer paper works well for constructing a timeline. If you cannot get this, you can tape sheets of plain paper together end to end. This timeline begins with Creation (10,000) then prehistory in 5,000 BC. Divide the timeline according to the intervals given below. Each division should be at least an inch. Each interval will represent a span of time. The spread of each interval is shown in the table below. For example, divide the intervals between 5,000 and 450 BC, into 100 years intervals.

From		To		Division (Yrs)
Creation	BC	10,000	BC	
5,000	BC	450	BC	100

Across the top of the timeline the Ages should be shown. The following chart shows the different Age divisions. Note that some of these overlap.

Age Divisions	From		To	
Prehistory	Prehistory	BC	3,500	BC
Ancient	3,500	BC	500	BC

The following is an example of a segment of the timeline. The subjects covered are listed down the leftmost column.

Middle East		Sumeria		- - - - - - - - - - - - - - - -
Asia				
Africa			Egypt	- - - - - - - - - - - - -
Mediterranean				
Creation	10,000	5,000	4,900	4,800	4,700
W. Europe				
E. Europe				
N. America				
S. America				

It's About Time, a timeline book is available through TRISMS for those who have limited wall space or take part in mobile classrooms. You will have to fill in the dates and topics.

You will notice there are variations in dates as well as names. We employ the ones that seem to be most frequently used. Dates listed with the civilizations are approximate and have been rounded. In many instances the dates fixed are the times encompassing the zenith of these civilizations, even though many have a history prior to these fixed dates. For the Middle School student we feel it is important that they are concerned only with the civilization in the height of its power and its subsequent decline. For the High School student you might want to dig into the civilization's prehistory before these dates and see how many of them interacted and borrowed from, or were influenced by each other.

COURSEBOOK

The student will create a coursebook on Civilizations. It will include all questionnaires for Civilization, Art, Music, and Architecture, also all assignments for Vocabulary, Science, Worksheets, Literature, maps, and any other interesting information on the subject. If they did *The History Makers* (TRISMS Volume One) previously they may combine the old coursebooks with the new.

The questionnaires and maps the student will use are located in the Student Pack.

Constructing the Coursebook

To make the coursebook you will need to obtain one, three-inch 3-ring notebook, construction paper or something similar for title pages and multiple copies of all questionnaires and maps. Start the coursebook with an identifying title page. Fill the notebook with alternating title and form pages. You can label the title page with the name and date of the civilization being studied. The questionnaire will precede any drawings, articles or pictures the student may find or photocopy. This is their own textbook and they should be encouraged to make it interesting to themselves.

Using the Coursebook Questionnaires

Questionnaires are provided in the Student Pack for each subject. The student will fill these out during the course of the year as each topic is studied. They should give a brief but thorough answer for each question. If more explanation is required to answer completely, put this information on a separate sheet of paper. Answers may vary depending on the reference material available. Be willing to accept an answer if it is accurate to the civilization or time period.

The following sections expound on the use and meaning of the questions asked on each questionnaire.

CIVILIZATION QUESTIONNAIRE

You will complete a questionnaire for every civilization you study. This questionnaire asks the following questions:

♦ What is the name of the civilization?

  How did they receive this name? Did they call themselves by this name or did their neighbors or enemies name them? In some cases, the archaeologist identifies them with a name. How did they select it?

♦ What is the time span of this civilization? (The dates we provide are not necessarily from their earliest history.) How many years did this civilization exist?

♦ When was the zenith of this civilization? How long did they maintain this level of excellence? What brought about the apex in this civilization?

♦ From what direction did these peoples migrate? Where did they live originally? Why did they relocate?

♦ What is the global location of this civilization? What present-day country and/or peninsula did they live on?

♦ What is the topography of the region this civilization occupies? This should be a detailed description of the region.

♦ What type of agriculture does this region or civilization cultivate? Did these people live off the land? What kinds of crops did they grow? Did they bring plants with them from their homeland? Did they develop new crops for the area?

♦ What animals are native to this region? Did this civilization domesticate these or did they bring new animals to this region?

♦ What form of government does this civilization function under? Ex. monarchy, theocracy, republic

♦ What type of laws governed this civilization? Give examples of the kinds of laws they had to obey.

  Ex. an eye for an eye, rights for citizens, penalties, retribution, etc.

  Who did the law protect or benefit?

♦ Who are some famous individuals from this civilization? These can be anybody. If they are not given a place in history books, explain why you feel they are famous. Many times it is the assistant who is the real hero or mastermind.

♦ What is the religion of this civilization? Many civilizations had more than one religion to follow. List the major religions of this civilization. What part did religion play in each civilization?

        Ex. Judaism, polytheism, animism, etc.

♦ What did the people believe?

>> Ex. one god, many gods, an afterlife, forgiveness, fate, superstitions, etc.

♦ How do their beliefs affect their lives?

Did they observe holy days, make offerings or sacrifices? Were they superstitious?

♦ Who were the religious leaders of the civilization?

>> Ex. shamans, heads of government, prophets, priests, etc.

♦ Give a physical description of a person from this civilization. Many civilizations were a mixed population; describe the conquerors. Also identify their race - Semitic, Indo-European, Aryan, etc.

♦ What type of visual arts is this civilization known for?

>> Ex. painting, sculpture, metalwork, pottery, mosaics, woodwork, etc.

♦ What types of music and dance were practiced in this civilization?

>> Ex. military bands, folk music and dance, liturgical, opera, ceremonial

♦ Did they express themselves through drama? What type of drama was written in this civilization?

>> Ex. comedy, tragedy, liturgical, etc.

♦ What type of literature did this civilization produce?

>> Ex. epics, myths, folktales, legends, poetry, etc.

♦ What kind of architecture is this civilization known for?

>> Ex. pyramids, palaces, yurts, etc.

♦ What advancements in science is this civilization noted for?

>> Ex. metal casting, surgery, medicine, agriculture

♦ What is the economic structure of this civilization? What is their economy based on?

>> Ex. trade, currency, barter system, agrarian, self-sufficient, etc.

♦ Explain their economic system

>> Ex. The Phoenician society was based on trade. Their people worked to develop trade goods. They set up trading posts in the Mediterranean Sea to market these items.

♦ What are this civilization's exports? These are crops or items that are produced in such abundance that they can be traded or sold to other countries.

♦ With whom did this civilization trade? What was the scope of their trade routes? What countries and peoples did they trade with? What kinds of things did they trade for?

♦ Who was educated in this civilization? Not every civilization allowed for education of the masses. Who were these privileged ones and what did they learn?

♦ How were they educated?

Who taught them? Ex. tutors, priest, parent, craftsman, etc.

Was there an educational institution? Ex. school, synagogue, boarding school, apprenticeship, etc.

What subjects were they taught?

♦ What was the written language of this civilization? Many civilizations have more than one language or dialect. Ex. Akkadian, Sanskirt, Hieroglyphics, etc.

What type of written language was it? Ex. pictograph, phonogram

♦ What did this civilization do for entertainment?

Ex. religious functions, storytelling, puppets, games, festivals, etc.

♦ What kinds of sports did they participate in?

Ex. Olympics, chariot racing, wrestling, ball games, acrobatics, etc.

♦ What was the social structure of this civilization? How were the social classes decided?

Ex. aristocrats, merchants, slaves, free men, etc.

♦ What role did women have in this civilization? What were their responsibilities? What part did they play in society, religion, ownership, government, business, home making?

♦ What was it like to be a child in this society? What were their roles and duties? Were boys treated differently than girls? Why?

♦ What accomplishment is this civilization known for? This accomplishment can be from any category.

Ex. invention, literature, communication, architecture, exploration, conquest, transportation, government, etc.

♦ Did this civilization borrow from an earlier civilization? Did these people take an idea and expand on it? Did they build on the technology or knowledge of previous peoples? Give examples.

♦ Did this society change slowly or quickly? You will find that some civilizations like Egypt changed very slowly. Others changed very quickly like the Mycenaeans. Explain why or how this occurred.

♦ What has become of this civilization? What happened to them?

Ex. Was it conquered, absorbed, exiled, destroyed or still exists.

CIVILIZATION SHORT FORM

Some civilizations left little written or deciphered history yet they play an important role in the development of nations. A Civilization Short Form will be used to study these. This questionnaire asks the following questions:

♦ What is the name of the civilization?

♦ What is the time span of this civilization? How many years did this civilization exist?

♦ When was the zenith of this civilization?

♦ From what direction did these peoples migrate? Where did they live originally?

♦ What is their present-day global location? (what country and/or peninsula)

♦ What is the topography of the region this civilization occupies?

♦ What type of agriculture does this region or civilization cultivate?

♦ What form of government does this civilization function under?

♦ Who are some famous individuals from the civilization?

♦ What are the religious beliefs of the people? How do these affect their lives?

♦ Who are the religious leaders in this civilization?

♦ What type of visual arts is this civilization known for?

♦ What types of music and dance were practiced in this civilization?

♦ What type of drama was written in this civilization? What types of literature did this civilization produce?

♦ What kind of architecture was this civilization known for?

♦ What advancements in science was this civilization noted for?

♦ What was the economic structure of this civilization?

♦ What did this civilization do for sports and entertainment?

♦ What is the social structure of this civilization?

♦ What accomplishment is this civilization known for?

♦ Did this civilization borrow from an earlier civilization?

♦ What has become of this civilization?

ART HISTORY QUESTIONNAIRE

The study of art for the Ancient world will be directed at specific pieces by which the civilization or culture is known. You will complete a questionnaire for every item you study. The questionnaire asks the following questions:

- What is the name of the item?
- What civilization did this item come from or was developed by?
- What was the functional use or purpose of the item? Is it suited to a particular operation or use? (ex. jewelry: ornamental accessory)   What is its practical or symbolic advantage or result (ex. symbolized wealth, often had significance of rank or as an amulet against evil)
- What inspired this art? (nature, animals, religious belief)
- In what locations are the works found? Where do archaeologists discover these items or where would they normally be found in the day-to-day life of the civilization?
- What trade regions were these items found in? (if information is available)
- What technique is used to make the item? These are the working methods or manner of performance.
       (ex. lost wax casting, potter's wheel, weaving)
- What materials are used to make this item? List the ingredients or materials used.
- How is this item made? List the steps to the process.
-     Draw the item or obtain a picture.

MUSIC HISTORY QUESTIONNAIRE

The music of the ancient civilizations will be studied with the following questionnaire.

- What is the name of the civilization?
- What types of instruments do these people play?

   Percussion -- drums, gong, cymbals

   Strings -- harps, koto, banjo, mandolin

   Wood winds -- oboe, clarinet

   Horns -- shells, animal horns, trumpets

- What materials were these instruments made from?   Ex. flute made from reeds, horn made from shell, rattles made from gourds, etc.
- What was the purpose of music in this civilization? Ex. liturgical, folk music, military band, drama
- Trivia: What did you find interesting or unusual about their music or instruments.
- Draw an example of an instrument used by this civilization or obtain a picture.

ARCHITECTURE QUESTIONNAIRE

You will complete an architecture questionnaire for every civilization studied. You will look at specific structures, as well as home construction and city planning. This questionnaire is in three parts and asks the following questions.

Part I

♦ What is the name of the specific structure being studied?

♦ What is the name of the civilization?

♦ Was this structure built for public or private use? Who could utilize this building? Ex. homes are private while theaters are public.

♦ What was this structure used for? How was it specifically utilized?

Ex. temple, market, library, mausoleum

♦ What types of materials were used in this building's construction? Geographic location can dictate what materials are available for building.

Ex. clay, quarried stone, lumber, bricks

♦ Were these materials man made or natural?

Ex. Natural - wood, stone, reeds, pitch, mud

Man made - brick, tiles, plaster

♦ Were special materials or tools developed to build this structure?

Ex. Tools - plumb line, boning rod, chisel, saw

♦ What type of machinery was used to construct this building?

Ex. inclined plane, wheel, lever

♦ What features are distinctive of this building style?

Ex. post & lintel, vault, arch, columns

Part II

♦ How are the homes arranged in this civilization? How do families live?

Ex. one room hut, tent, two-story dwelling, villa with courtyard

♦ What were their homes constructed of? What kinds of building materials are used in-home construction?

Ex. mud & wattle, thatch, bricks, lumber, stone

Part III

♦ How were the towns arranged by this civilization? Was a plan used to place the streets, homes, businesses, and temple? How did they choose the location to build on? Was it based on good land and water resources, easy fortification, or availability of building materials?

♦ What was the focus or center of town? What was the focal point of these people?

      Ex. temple, citadel, palace, market

♦ What sorts of defenses were used? How did these people protect themselves? Ex. walls, sea, cliffs, towers

ALL STUDENTS

All students may want to develop a special interest coursebook or project. Most students will find something that interests them but find there is not enough information covered in the unit plan. Examples include: food--make a cookbook; sports before Nike; boats--build models; weapons and military strategy; language and communications; fashion--hair and clothes styles; history of fabric and fasteners; flags; guilds; entertainment--learn to juggle, make their game boards and play their games; drama--write and act out a play, folk dancing, myths, and folklore. Whatever strikes your fancy! Students should include their special interest on their timeline. They can design their own information questionnaire.

RESEARCH GUIDE

One of the goals of TRISMS is to teach research skills, to empower the student to become an independent learner. This is accomplished by using a directed study as a framework and coursebook development as an end product. The Resource and Reading Section lists books we have found helpful in answering the questionnaires. However, your library may not carry these titles. The encyclopedia is a good place to start. You may start with a specific search, such as Sumerians. If the subject is not found you may have to broaden the search to include geography or time frame to be studied, such as Ancient Mesopotamia. As you read, keep in mind the questions you are trying to answer. When using the encyclopedia be aware of cross-references. These provide the reader with more information on the topic. Readers are directed to other articles by the words *see, see also,* or *see under* followed by the subject heading. The abbreviation (q.v.) stands for the Latin words "quod vide", meaning *which see*. When this abbreviation follows a word or name, it signals that the word itself is the title of a separate article in the encyclopedia. The index helps you determine if there is information on your subject included within the text. The supplementary bibliography lists other sources outside of the encyclopedia.

It is unlikely that one book will have all the answers you need to complete a questionnaire and worksheet. You can quickly evaluate books by scanning the table of contents or index. If the table of contents does not list your subject it is unlikely to be in the book. You can look up keywords in the index to see if there is any helpful information within the book on your topic. The bibliography may direct you to other books on your topic.

The internet can be a valuable research tool. There is an internet tutorial on the TRISMS site as well as Homework Helps listing sites discovered by families using TRISMS.

You will need to sharpen your map skills and compare ancient and modern maps. A good atlas is a must.

Webster's *Encyclopedia of Dictionaries* is a wonderful resource for some of those vocabulary words.

REFERENCE AND RESEARCH HELPS LISTED

♦ Library
♦ Card Catalogue (microfiche or computer)
♦ Library of Congress
♦ Dewey Decimal system
♦ Call Numbers
♦ Dictionary
♦ Expository Dictionary
♦ Encyclopedia
♦ Atlas
♦ Almanac
♦ Periodicals
♦ Reader's Guide to Periodical Literature
♦ Reference Books
♦ Fiction
♦ Nonfiction
♦ Biography
♦ Bibliography
♦ Thesaurus
♦ Index
♦ Glossary
♦ Table of Contents
♦ TRISMS Homework Helps

QUIZZES AND TESTS

All quizzes and tests are located together in the Student Pack. Vocabulary Quizzes and Unit Quizzes are given per unit and should be used to help study for the Semester Test. There is a separate test for Vocabulary, Literature, Humanities, and Civilizations. The teacher should make sure the student covers in there study everything included in the test. Students should study all worksheets, vocabulary words, questionnaires, maps, as well as any information presented with the Literature and Science assignments.

ANSWER KEYS

Answer keys will be located in the Answer Key book. Answers may vary. You should be willing to accept an answer if it is accurate to the civilization or time period. Students must be able to justify other answers from their research and coursebook.

If the student is having difficulty finding information the key may be used as a resource. Don't forget to go to Homework Helps at www.trisms.com.

NECESSARY ADDITIONAL RESOURCES

Necessary resources include an atlas, globe, world map, encyclopedia set, thesaurus, dictionary, and access to a public library or internet.

You will need to utilize the world map weekly. It is essential to examine the extent of exploration and vital to understanding the origins of civilizations and their expansion.

1 - 3 inch three-ring notebook

1 - 3x5 or 4x6 card file box (for younger students reading historical fiction and making mini reports)

1 - ruled 3x5 or 4x6 index cards (for mini reports)

1 - set of colored pencils, markers or crayons

Construction paper or similar weight paper for coursebook title pages

Ruled notebook paper

Coiled index cards for vocabulary study guide

Access to a copy machine for extra pictures, maps, etc.

USING TRISMS FOR MULTI-LEVEL TEACHING

TRISMS *Discovering the Ancient World* builds on what was learned in *The History Makers* (TRISMS Volume One). We begin with the understanding that the student has a chronological view of history, knows how to research, and has basic writing skills. If you desire to adapt the curriculum for younger students you will need to simplify the curriculum. This can be done by using the Civilization Short Form questionnaires. Eliminate the weekly compare question requiring abstract reasoning. Use a product like Daily Grams for grammar requirements. Literature is important to understanding the civilization. We suggest the student read the literature but replace the assignment with an exercise asking and answering Who, What, When, Where, and How. You will need to pick reading material that is commensurate with the student's reading level, historical fiction and junior level resource books are provided in the Resource and Reading List.

MINI-REPORTS

The younger student will make mini-reports on all historical fiction and biographies read during the year. These reports should be made on 3x5 or 4x6 index cards. The format is as follows:

♦ Place the title of the book on the top line.
♦ Capitalize the first, last and all-important words in the title of the book.
♦ The student should always underline the title.
♦ On the next line write the name of the author of the book.
♦ Skip a line and write a few sentences about the story. Make it interesting without giving away the ending.
♦ In the bottom, left-hand corner record the number of pages in the book.
♦ In the bottom right corner write the ISBN number.

Students will keep their mini reports in a card file box.

SIXTH GRADE

The sixth grade student should study the listed civilizations using the Civilization Short Form; worksheets are optional. Junior level resource books are listed in the Resource and Reading List. Sixth graders may skip unit plan items listed under Historical Events, Other Areas of Interest, and the Compare Question. The sixth grade student can read the Literature and answer the Who, What, When, Where, and How questions. The literature and composition assignments may be challenging, therefore historical fiction is acceptable reading. A mini report should be written for each historical fiction or biography read.

SEVENTH GRADE

The seventh grade student should study the listed information in the unit plans. They may choose to use the Civilization Short Form; worksheets are optional. The literature should be read to aid understanding of the civilization, then answer the Who, What, When, Where and How questions to check comprehension. Let the student try the composition assignments but they will be challenging. The Compare Question is optional. You could fill in the timeline with the items listed under Historical Events in the unit plans. Discuss the impact of these events on the people studied. Use a grammar supplement.

EIGHTH GRADE

The eighth grade student should follow the given information in the unit plans. In addition, they should include a grammar supplement.

HIGH SCHOOL

The following is a list of possible high school credits that can be claimed after completing TRISMS *Discovering the Ancient World*. Electives may be half or whole credits depending on how much time you spend. Whole credits are 120 hours or a ten-page research paper. Include your special interests subject as an elective.

Full credit	Electives
World or Ancient History	Biblical History
Literature & Composition	Multi-cultural studies
or English I	Anthropology
Survey of Science History or General Science	Music History
World or Ancient Geography	Art History
	Research Writing (1/2 credit)
	Critical Thinking I
	Cartography (1/2 credit)
	Historical Architecture
	Special interest topic

For those interested in an Economics credit, study the civilization and write a paragraph on how each civilization taxed its people and how the revenue was used. Answer the question, "Did this form of taxation stimulate the economy or did it contribute to its demise?" Explain your answer and support it with evidence and logic. *For Good and Evil - The Impact of Taxes on the Course of Civilization* by Charles Adams will be a helpful resource.

Resource &
Reading List

<u>Resource and Reading List</u> is divided into the following sections.
- Questionnaire resource books -- R
- Young Adult -- YA
- Junior level resource books -- J
- Historical Fiction -- HF
- Picture books -- E
- Periodicals -- P
- Video tapes -- V
- Audio tapes -- A

These books will aid the teacher as well as the student. For younger students we suggest using books from the youth section because they give the main idea along with pictures and helpful diagrams. The periodicals, video and audio tapes are optional but can enhance the learning process.

The books and films listed in the unit plan should be available from a public library. Several alternative book selections are listed in order to provide variety and facilitate availability. Books may be substituted as long as they fit the topic and time period.

The books and films listed are to help the student find information. However, if you are unable to locate listed books your librarian is your best resource. She is there to help you. Tell her the subject or time period you are looking for, she'll be glad to locate a book for you. Or you can search the library catalog system yourself. Look under subject, time period, or key word. All books will be catalogued as biography or fiction; otherwise they are a type of reference book.   Remember to use the Dewey Decimal system. Consider using the internet, there is a search engine tutorial at www.trisms.com as well as Homework Helps.

<u>Helpful books for many units:</u>
Adams, Charles. *For Good and Evil - The Impact of Taxes on the Course of Civilization*. NY: Madison Books, 1993.

Davison, Michael Worth, Ed. *Everyday Life through the Ages*. London: Reader's Digest Association Limited, 1992.

Diagram Group. *Musical Instruments of the World - An Illustrated Encyclopedia*. NY: Fact on File Publication, 1976.

Durant, Will and Ariel. *Our Oriental Heritage*. NY: Simon & Schuster, 1935 - 1975.

Geography Dept. *Scrawl! Writing in Ancient Times*. MN: Runestone Press, 1994.
Buried Worlds series

Janson, H. W. *History of Art*. NY: Abrams, 1991.

The Kingfisher *Illustrated History of the World - 40,000 B.C. to Present Day*. NY: Kingfisher Books, 1992.

Leon, Vicki. *Uppity Women of Ancient Times.* NY: MJF Books. 1995. Just for Fun!!

Leithart, Peter J. *A House for My Name: A Survey of the Old Testament.* Moscow, ID: Canon Press, 2000.

Leithart, Peter J. *Heroes of the City of Man.* Moscow, ID: Canon Press, 1999.

Somerset Fry, Plantagenet. *The Dorling Kindersley History of the World.* NY: DK Publishing House, 1994.

Stanton, Mary, & Hyma, Albert. *Streams of Civilization - Volume One.* IL: Christian Liberty Press, 1976.

Time-Life Books, Ed. *TimeFrame Books* series. Alexandria, VA

Unit 1   In The Beginning

R   *The Bible*

P   "The Bible", <u>Life Magazine</u>, Vol. 57, No. 26 12/25/64 double issue.

HF   Beechick, Ruth. *Adam and His Kin: The Lost History of Their Lives and Times.* Arrow Press, 1991.

J   Caselli, Giovanni. *The First Civilizations.* NY: Peter Bedrick Books, 1985.
        *evolution included

HF   Denzel, Justin F. *Boy of the Painted Cave.* NY: Philomel Books, 1988.
J   Tao, a young cave boy dreams of becoming a cave painter.

E   Gauch, Patricia Lee. *Noah.* NY: Philomel Books, 1994.

J   Glubock, Shirley. *Art and Archaeology.* NY: Harper and Row, 1966.

R   Gombrich, E. H. *The Story of Art.* Englewood Cliffs, NJ: Prentice-Hall, 1984.

J   Martin, Ana. *Prehistoric Stone Monuments.* Chicago: Children's Press, 1993.
        World Heritage series

J   Oliphant, Margaret. *The Earliest Civilizations.* NY: Facts on File, 1993.
        * evolution included

R   Pfeiffer, Charles F. *Old Testament History.* Grand Rapids, MI: Baker Book House, 1973.

HF   Pryor, Bonnie. *Seth of the Lion People.* NY: William Morrow Jr. Books, 1988.
J   Seth, a young man with a twisted leg is tolerated by his people. He must find a way to be their leader.

J   Terzi, Marinella. *Prehistoric Rock Art.* Chicago: Children's Press, 1992.

R   Time-Life Books, Ed. *The World's Last Mysteries*. Alexandria, VA: Time-Life Books, 1978.

V   The Bible....In the Beginning. Videocassette. Dir. John Houston. 20th Century Fox, 1966. 171 min.

         This video shows the life of hunters and gatherers in the Kalahari Desert in Africa.
V   The Bushman of the Kalahari, National Geographic Video
         Be reminded that National Geographic's policy is to show people as they are.

V   Ancient Mesopotamia. Videocassette. Dir. Coronet Instructional Media. Coronet Films and Video. 1976.
         10 min.

V   Mesopotamia - Return to Eden. Videocassette. Dir. Robert Gardner. Time-life Video & Television, 1995,
         48 min. Time Life's Lost Civilizations series.

Unit 2   Sumerians

R   *Civilizations of the Middle East*. Austin, TX: Raintree Steck-Vaughn Library, 1992.
         History of the World series

R   Coblence, Jean-Michael. *The Earliest Cities*. Morristown, NJ: Silver Burdett, 1987.

R   Cottrell, Leonard. *The Quest for Sumer*. NY: Putman, 1965.

R   Durant, Will and Ariel. *Our Oriental Heritage*. NY: Simon & Schuster, 1935 - 1975.

R   *Epic of Gilgamesh*. Trans. N.K. Sandars. NY: Amereon House, 1990.

R   Foster, Leila. *The Sumerian*. NY: Franklin Watts, 1990.

P   Heyerdahl, Thor, "The Tigris Expedition", National Geographic, 12/1978.

J   Odijk, Pamela. *The Ancient World - The Sumerians*. Englewood Cliffs, NJ: Silver Burdett Press, 1990.

R   Time-Life, Ed. *Sumer: Cities of Eden*. Alexandria, VA: Time Life Books, 1993.

R   Unstead, R. J. *Looking at Ancient History*. New York: Macmillan, 1960.

R   Whitehouse, Ruth. *The First Cities*. Phaidon, NY: E.P. Dutton, 1977.

V   Abraham. Videocassette. Dir. Raffelle Mertes. Turner Pictures, 1994. 150 min.

Unit 2   Indus Valley

R   Edwardes, Michael. *Indian Temples and Palaces.* Feltham, NY: Hamlyn, 1969.

P   Edwards, Mike, "Indus Civilization-Clues to an Ancient Puzzle", National Geographic, 6/2000.

J   Ellis, John, and Neurath, Marie. *They Lived Like This in Ancient India.* NY: Franklin Watts, 1967.

R   Fairley, Jean. *Lion River the Indus.* NY: John Day Co., 1975.

J   Galbraith, Catherine, and Mehta, Rama. *India Now and Through Time.* Boston: Houghton Mifflin, 1980.

R   Kramrisch, Stella. *The Art of India.* Delhi: Motilal Banarsidass, 1987.

J   Mehta, Rama, and Galbraith, Catherine. *India Now and Through Time.* Boston: Houghton Mifflin, 1980.

R   Possehl, Gregory L. *Indus Age: The Writing System.* Philadelphia: University of Pennsylvania Press, 1996.

R   Possehl, Gregory L. *Indus Age: The Beginnings.* Philadelphia: University of Pennsylvania Press, 1999.

R   Time Life, Ed. *Ancient India: Land of Mystery.* Alexandria, VA: Time-Life Books, 1994.

R   Unstead, R.J. *How They Lived in Cities Long Ago.* Ed. Adrian Sington, NY: Arco Pub., 1981.

R   Wheeler, Sir Mortimer. *Civilizations of the Indus Valley and Beyond.* NY: McGraw-Hill, 1966.

J   Wilkinson, Philip, and Pollard, Michael. *Mysterious Places - The Magical East.* NY: Chelsea House Pub., 1994.

R   Whitehouse, Ruth. *The First Cities.* Phaidon, NY: E.P. Dutton, 1977.

         This film covers the Indus Valley, Asoka, and Mauyran periods.
V   India: The Empire of the Spirit. Videocassette. Produced by Maryland Public Television and Central Independent Television, U.K. Ambrose Video, Program 2, 1991. 60 min.
         *nudity in the last five minutes (ceremonial temple run to river)

V   Jacob. Videocassette. Dir. E. Guaneiri. Turner Pictures, 1994. 94 min.

Units 3 & 4   Egypt

J   Allen, Kenneth. *One Day in Tutankhamen's Egypt.* NY: Abelard-Schuman, 1974.

E   Angeletti, Roberta. *Nefertari, Princess of Egypt.* NY: Oxford University Press, 1998.

TRISMS©

P   Arden, Harvey, "In Search of Moses", <u>National Geographic</u>, 1/1976.

HF   Berry, Erick. *Honey of the Nile*. NY: The Viking Press, 1963.
J   After the death of Ikhnaton, Egypt is on the verge of civil war.

HF   Carter, Dorothy Sharp. *His Majesty, Queen Hatshepsut*. NY: J.B. Lippincott, 1987.

J   Chadefaud, Catherine. *The First Empires*. Morristown, NJ: Silver Burdett, 1988.
         The Human Story series.   *some nudity

HF   Chubb, Mary. *Nefertiti Lived Here*. NY: Thomas Y. Crowell Co., 1954.
Y   Story of an archaeological dig at Armana during 1930.

J   Crosher, Judith. *Technology in the Time of Ancient Egypt*. Austin, TX: Raintree Steck-Vaughn., 1998.

J   Crosher, Judith. *Ancient Egypt*. NY: Viking, 1993.

J   Cross, Wilber. *Egypt*. Chicago: Children's Press, 1982.

J   David, A. Rosalie. *The Egyptian Kingdoms*. NY: Peter Bedrick Books, 1988.
         The Making of the Past series.

J   David, Rosalie. *Growing Up in Ancient Egypt*. Troll Associates, 1994.

P   Dothan, Trude, "Lost Outpost of the Egyptian Empire", <u>National Geographic</u>, 9/1913.

R   Durant, Will. *Our Oriental Heritage*. NY: Simon & Schuster, 1935 - 1975.

J   Erman, A. *The Literature of the Ancient Egyptians*. NY: Dutton, 1927.

J   Glubok, Shirley, and Tamarin. *The Mummy of Ramose - The Life and Death of an Ancient Egyptian*. NY: Harper & Row Pub., 1978.

J   Glubok, Shirley. *The Art of Egypt Under the Pharaohs*. NY: Macmillan, 1988.

J   Glubok, Shirley. *The Art of Ancient Egypt*. NY: Athenium, 1962.

E   Grant, Joan. *The Monster that Grew Small*. NY: Lothrop, Lee and Shepard Books, 1987.
         Egyptian folklore in which a timid boy finds courage.

J   Harris, Geraldine. *Gods and Pharaohs from Egyptian Mythology*. NY: Peter Bedrick Books, 1982.

HF   Henty, G. A. *The Cat of Bubastes - A Tale of Ancient Egypt*. London: Blackie and Son, 1888. PA: Preston Speed, 1998.

HF   Jones, Ruth Fosdick. *Boy of the Pyramids*. NY: Random House, 1952.
J   Khufu's Egypt

J   Leacroft, Helen, and Richard. *The Buildings of Ancient Egypt*. NY: W. R. Scott, 1963.

HF   McGraw, Eloise. *Mara, Daughter of the Nile*. NY: Puffin Books, 1985.
YA   A slave girl must find a way to serve two masters.

HF   McGraw, Eloise. *The Golden Goblet*. NY: Puffin Books, 1986.
J   An Egyptian boy who wants to become a goldsmith.

HF   Meadowcroft, Enid LaMonte. *The Gift of the River*. NY: Tomas Y. Crowell Co., 1975.
J   a story history of Egypt

R   Metz, Barbara. *Temples, Tombs and Hierglyphs*. NY: Dodd, Mead, 1978.

J   Morley, Jacqueline. *So You want to be an Ancient Egyptian Princess?*. Brookfield, CN: Twenty-First Century Books, 1999.

HF   Morrison, Lucile. *The Lost Queen of Egypt*. NY: Stokes, 1937.
        Egyptian life during Akhenaten's reign

J   Raphael, Elaine, and Bolognese, Don. *Drawing History - Ancient Egypt*. NY: Franklin Watts, 1989.

HF   Rubalcaba, Jill. *A Place in the Sun*. NY: Clarion Books, 1997.
J   13th c. B.C. Senmut, the gifted son of a sculptor is taken into slavery but his talent rescues him.

E   Sabuda, Robert. *Tutankhamen's Gift*. NY: Atheneum Books for Young Readers, Simon & Schuster, 1999.

J   Sandak, Cass R., and Purdy, Susan. *Ancient Egypt*. NY: Franklin Watts, 1982.

J   Schlein, Miriam. *I, Tut the Boy Who Became Pharaoh*. NY: Four Winds Press, 1979.

P   Smith, Ray Wilfrield, "Computer Helps Scholars Re-create an Egyptian Temple", (Akhenaten Temple Project), National Geographic, 11/1970.

R   Stewart, Desmond. *The Pyramids and Sphinx*. NY: Newsweek Book division, 1971.

E   Stolz, Mary. *Zekmet The Stone Carver: A Tale of Ancient Egypt*. NY: Harcourt Brace Jovanovich, Publishers, 1988.

J   Time-Life Books, Ed. *What Life Was Like on the Banks of the Nile - Egypt 3030 - 30 B.C.* VA: Time-Life Inc., 1997.   *What Life Was Like* series.

R   White, J. E. Manchip. *Everyday Life in Ancient Egypt.* Westport, CN: Greenwood Press, 1974.

J   Woods, Geraldine. *Science in Ancient Egypt.* NY: Franklin Watts, 1988.

V   Egypt: Quest for Eternity. Videocassette. Dir. Norris Brock. National Geographic Society Video, 1982. 60 min.

V   Egypt Secrets of the Pharaohs. Videocassette. Dir. Carl Andrew. National Geographic Society Video, 1997. 60 min.

V   This Old Pyramid. Videocassette. Dir. Michael Barnes. Nova, 1992. 60 min.

V   Joseph. Videocassette. Dir. Roger Young. Turner Home Entertainment, 1990's, 185 min. PG-13.

Unit 5 Phoenicians

R   *Civilizations of the Middle East .* Austin, TX: Raintree Steck-Vaughn Library, 1992.
        History of the World series

R   Edey, Maitland A. *The Sea Traders.* NY: Time-Life Books, 1974.

R   Herm, Gerhard. *The Phoenicians - The Purple Empire of the Ancient World.* NY: Morrow, 1975.

HF   Flaubert, Gustave. *Salambo.* Trans. Powys Mathers. London: Folio Society, 1940.
A   ancient Carthage

P   Matthews, Samuel, "Sea Lords of Antiquity", National Geographic, 8/1974.

J   Odijk, Pamela. *The Phoenicians.* Englewood Cliffs, NJ: Silver Burdett Press, 1989.

J   Simon, Charnan. *Explorers of the Ancient World.* Chicago: Children's Press, 1990.

R   Time Life, Ed. *Barbarian Tales.* Alexandria, VA: Time-Life Books, 1987.

Unit 6 Minoans

J   Asimov, Isaac. *How Did We Find Out About Earthquakes?.* NY: Walker, 1978.

TRISMS©

J   Caselli, Giovanni. *In Search of Knossos: The Quest for the Minotaur's Labyrinth.* NY: Peter Bedrick Books. 1999.

J   Ceserani, Paolo, and Ventura, Piero. *In Search of Ancient Crete.* Morristown, NJ: Silver Burdett, 1985.

R   Cotterell, Arthur. *The Minoan World.* NY: Scribner, 1980.

R   Cottrell, Leonard. *Crete: Island of Mystery.* Englewood Cliffs, NJ: Prentice-Hall, 1965.

R   Cottrell, Leonard. *The Mystery of Minoan Civilization.* NY: World Pub. Co., 1971.

R   Cottrell, Leonard. *The Bull of Minos: The Discoveries of Schliemann and Evans.* NY: Facts on File, 1984.

J   Edmond, I. G. *The Mysteries of Homer's Greeks.* NY: Elsevier: Nelson Books, 1981.

HF   Faulkner, Nancy. *The Traitor Queen.* NY: Doubleday, 1963.
J   Crete during the reign of King Minos

J   Higgins, Reynold. *The Archaeology of Minoan Crete.* NY: H. Z. Walck, 1975.

R   Higgins, Reynold. *Minoan and Mycenaean Art.* NY: Oxford University Press, 1981.

P   Judge, Joseph, "Minoans and Mycenaeans: Greece's Brilliant Bronze Age", National Geographic, 2/1978.

E   Lattimore, Deborah Nourse. *The Prince and the Golden Ax - A Minoan Tale.* NY: Harper & Row, 1988.

P   Marinatos, Spyidon, "Thera, Key to the Riddle of Minos", National Geographic, 5/1972.

R   MacGillivray, Joseph Alexander. *Minotaur: Sir Arthur Evans and the Archaeology of the Minoan Myth.* NY: Hilland Wang, 2000.

HF   Palmer, Myron Tim. *Treachery in Crete.* Boston: Houghton Mifflin, 1961.
J   Two Egyptian boys discover a different lifestyle and intrigue on Crete.

J   Simon, Chaman. *Explorers of the Ancient World.* Chicago: Children's Press, 1990.

P   Stillwell, Agnes, "Where Sea-Kings Reigned", National Geographic, 11/1943.

R   Time Life, Ed. *Barbarian Tides -Time Frame 1500 - 600 B.C.* Alexandria, VA: Time-Life Books, 1987.

Y   Warren, Peter. *The Aegean Civilizations.* NY: Peter Bedrick Books, 1989.
      The Making of the Past series   *Palaeolithic period discussed in the book.

J   Wilkinson, Philip, and Dinee, Jacqueline. *Mysterious Places - The Mediterranean.* NY: Chelsea House Pub., 1994.

R   Willetts, R. F. *Everyday Life in Ancient Crete*. London: Batsford, NY: Putnam, 1969.

V   <u>Aegean - Legacy of Atlantis</u>. Videocassette. Time-Life Video and Television, 1995. 48 min., Time Life's Lost Civilizations series.

V   <u>Ancient Aegean</u>, Videosassette. Dir. JWM Productions. JWM Productions, 1998. 23 min, Ancient Civilizations for Children series. Ages 8-12.

Unit 7   Ancient China - Shang and Chou

R   Allan, Sarah. *The Shape of the Turtle: Myth, Art and Cosmos in Early China*. Albany, NY: State University of New York Press, 1991.   This book has good pictures of Chinese bronze work.

J   Beshore, George. *Science in Ancient China*. NY: Franklin Watts, 1988.

J   Chrisman, Arthur Bowie. *Shen of the Sea - Chinese Stories for Children*. NY: Dutton Children's Books, 1968.

J   Coblence, Jean-Michael. *Asian Civilizations*. Englewood Cliffs, NJ: Silver Burdett Press, 1988.

E   Demi. *The Emperor's New Clothes*. NY: Margaret K. Mc Elderry Book, 2000.
        A vain emperor is tricked into wearing magic clothes.

E   Demi. *Happy New Year! Kung-his-ts' ai'!*. NY: Crown Publishers, Inc., 1997.
        Examines the customs, traditions, foods, and lore associated with the celebration of Chinese New Year.

E   Demi. *Liang and the Magic Paint Brush*. NY: Henry Holt & Co., 1980.
        A poor boy receives a magic paintbrush that brings whatever he paints to life.

E   Demi. *The Magic Tapestry*. NY: Holt, 1994.

Y   Fang, Linda. *The Chi Lin Purse - A Collection of Nine Ancient Chinese Stories*. NY: Farrar, Straus, and Giroux, 1995.

J   *Favorite Children's Stories from China*. Beijing, China, Foreign Language Press, 1983.

J   Glubok, Shirley. *The Art of China*. NY: Macmillan, 1973.

P   Hyde, Nina, "Silk - The Queen of Textiles", <u>National Geographic</u>, 1/1984.

J   Kan, Lai Po. *The Ancient Chinese*. Morristown, NJ: Silver Burdett, 1980.

J   Knox, Robert. *Ancient China.* NY: Warwick Press, 1979.

E   Leaf, Margaret. *Eyes of the Dragon.* NY: Lothrop, Lee & Shepard Books, 1987.

E   Miller, Moira. *The Moon Dragon.* NY: Dial Books, 1989.

J   Martell, Hazel. *The Ancient Chinese.* NY: New Discovery, 1993.
        Worlds of the Past series

E   Mui, Shan. *Seven Magic Orders.* Ed. Ruth Tabrah, NY: Weatherhill, an Island Heritage Book, 1972.

R   Rodzinski, Witold. *The Walled Kingdom.* NY: Free Press, 1984.

R   Seeger, Eliazbeth. *Pageant of Chinese History.* NY: D. McKay Co., 1962.

J   Terzi, Marinella. *The Chinese Empire.* Chicago: Children's Press, 1992.
        World Heritage series

J   Wilker, Josh. *Confucius: Philosopher and Teacher.* NY: Franklin Watts, 1999.

E   Wolkstein, Diane. *White Wave.* NY: Thomas Y. Crowell, 1979.

E   Yolen, Jane. *The Emperor and the Kite.* NY: Philomel Books, 1988.

E   Young, Ed. *Lon Po Po.* NY: Philomel Books, 1989.

E   Young, Ed. *The Lost Horse.* NY: Harcourt Brace & Co., 1998.

E   Young, Ed. *Mouse Match.* NY: Silver Whistle, Harcourt Brace & Co., 1997.

E   Ziner, Feenie. *Cricket Boy.* NY: Doubleday & Co. Inc., 1977.

V   Ancient China. Videocassette. Dir. JWM Productions. Schlessinger Media, 1998. 23 min., Ancient
        Civilizations for Children series ages 8-12.

V   China - Dynasties of Power. Dir. Steve Eder. Time-life Video & Television, 1995. 48 min. Time-life's Lost
        Civilizations series.

Unit 8 Hittites

R   Brandon, S.G.F. *Ancient Empires.* NY: Newsweek, 1970.

R   Ceram, C. W. *Secret of the Hittites; the Discovery of an Ancient Empire.* NY: Knopf, 1956.

P   Dodd, Isabel F., "An Ancient Capitol", <u>National Geographic</u>, 2/1910.

R   Gurney, O. R. *The Hittites*. NY: Penguin Books, 1990.

P   Ramsay, Sir William, "A Sketch of the Geographical History of Asia Minor", <u>National Geographic</u>, 11/1922.

R   Roux, George. *Ancient Iraq*. Cleveland: World Pub. Co., 1965.

R   Wilkinson, Philip, and Dineen, Jacqueline. *Mysterious Places - The Lands of the Bible*. NY: Chelsea House Pub., 1994.

HF   Williamson, Joanne. *Hittite Warrior*. NY: Alfred A. Knopf, 1960.
        Based on a story from the Book of Judges

R   Unstead, R.J. *Looking at Ancient History*. NY: Macmillan, 1960.

Unit 8 Aryan India

J   Coblence, Jean-Michael. *Asian Civilizations*. Englewood Cliffs, NJ: Silver Burdett Press, 1988.

J   Ellis, John, and Neurath, Marie. *They Lived Like This in Ancient India*. NY: Franklin Watts, 1967.

R   Fairservis, Walter Ashlin. *The Roots of Ancient India; The Archaeology of Early Indian Civilization*. NY: Macmillan, 1971.

J   Mehta, Rama, and Galbraith, Catherine. *India Now and Through Time*. Boston: Houghton Mifflin, 1980.

J   Oliphant, Margaret. *The Earliest Civilizations*. NY: Facts on File, 1993.
        *evolution included

R   Thapar, Romila. *A History of India 1000 B.C. - A.D. 1526*. Penguin Books, 1966, 1990.

E   Shepard, Aaron. *Savitri - A Tale of Ancient India*. Morton Grove, IL: Albert Whitman and Co., 1992.

Unit 9 Old Babylon

R   Hillyer, Virgil, Mores. *Ancient World*. NY: McGraw-Hill, 1966.

J   Mellersh, H.E.L. *Sumer and Bablyon*. NY: Crowell, 1965.

R   Oates, Joan. *Babylon*. London: Thames & Hudson, 1979.

R   Roux, Georges. *Ancient Iraq*. Cleveland: World Pub. Co., 1965.

R   Saggs, H.W.F. *Everyday Life in Babylon and Assyria*. NY: Norton, 1966.

R   Unstead, R.J. *Looking at Ancient History*. NY: Macmillan, 1960.

J   Wilson, Kax. *A History of Textiles*. Boulder, CO: Westview Press, 1982.

Special Projects   Mysteries of the World

J   Branley, Franklyn. *The Mystery of Stonehenge*. NY: T.Y. Crowell Co., 1969.

P   Buxton, Going, "The Mysterious Prehistoric Monuments of Brittany", National Geographic, 7/1923.

J   Mann, Peggy. *Easter Island: Land of Mysteries*. NY: Holt, Rinehart, and Wilson, 1976.

R   National Geographic, Ed. *Mysteries of the Ancient World*. Washington: National Geographic Society, 1979.

R   *The World's Last Mysteries*. Pleasantville, NY: Reader's Digest Assoc., 1978.

V   Legends of Easter Island. Videocassette. Dir. Nova. Coronet Film & Video, 1989. 58 min.

Unit 10   Mycenaeans

J   Caselli, Giovanni. *In Search of Troy: One Man's Quest for Homer's Fabled* City. NY: Peter Bedrick Books, 1999.

HF   Coolidge, Olivia E. *The King of Men*. Boston: Houghton Mifflin, 1993.
based on the Agamemnon legend

R   Cottrell, Leonard. *Realms of Gold: A Journey in Search of the Mycenaeans*. Greenwich, CN: New York Graphic Society, 1963.

J   Edmonds, I.G. *The Mysteries of Homer's Greeks*. NY: Elsevior/Nelson Books, 1981.

R   Higgins, Reynold. *Minoan and Mycenaean Art*. NY: Praeger, 1967.

HF   Sutcliff, Rosemary. *Black Ships Before Troy*. NY: Delacorte, 1993.
Retells the Trojan War

J   Warren, Peter. *The Aegean Civilization*. NY: Peter Bedrick, 1989.
The Making of the Past * Paleolithic period discussed

V   <u>Aegean - Legacy of Atlantis</u>. Videocassette. Time-Life Video and Television, 1995. 48 min., Time Life's Lost Civilizations series.

Unit 11   Olmec

R   Bernal, Ignacio. *Olmec World*. Berkeley: University of California Press, 1969.

R   Coe, Michael. *America's First Civilization-Discovering the Olmec*. Princeton, NJ: Van Nostrand, 1968.

J   Glubok, Shirley. *The Art of Ancient Mexico*. NY: Harper and Row, 1968.

R   Reader's Digest, Ed. *Mysteries of the Ancient Americas - The New World Before Columbus*. Pleasantville, NY: Reader's Digest Association, 1986.

J   Stein, Conrad. *Mexico*. Chicago: Children's Press, 1984.
         Enchantment of the World series

P   Stuart, G. E.,"New Light on the Olmecs", <u>National Geographic,</u> Vol. 185, No. 5, 11/93.

J   Swanson, Earl Herbert. *The Ancient Americas*. NY: Peter Bedrick Books, 1989.

Unit 11   Nubia

J   Bianchi, Robert Steven. *The Nubians - People of the Ancient Nile*. Brookfield, CN: Millbrook Press, 1994.

HF   Bradshaw, Gillian. *The Dragon and the Thief*. NY: Greenwillow Books, 1991.
Y   fantasy

HF   Bradshaw, Gillian. *The Land of Gold*. NY: Greenwillow Books, 1992.
Y   fantasy

J   Corwin, Judith Hoffman. *African Crafts*. NY: Franklin Watts, 1990.

R   Goldston, Robert. *The Negro Revolution*. NY: Macmillan, 1968.

P   Gryzmski, K., "Nubia: Rediscovering African Kingdoms", <u>American Visions</u>, Vol. 8,
         O/N 1993.

J   Jenkins, Ernestine. *A Glorious Past - Ancient Egypt, Ethiopia and Nubia*. NY: Chelsea House Pub., 1995.

P   Kendall, Timothy, "Discoveries at Sudan's Sacred Mountain, Kingdom of Kush", <u>National Geographic</u>, 1990.

R   Quirke, Stephen, and Spencer, Jeffrey. *The British Museum Book of Ancient Egypt*. NY: Thames and Hudson, 1992.

P   Roberts, David, "Out of Africa: The Superb Artwork of Ancient Nubia", <u>Smithsonian</u>, 6/1993.

Units 12 & 13   The Hebrews

The Bible

R   Cheilk, Michael. *Ancient History*. NY: Harper Perennial, 1991.

J   Due, Andrea. *The Atlas of the Bible Lands*. NY: Peter Bedrick Books, 1998.

R   Durant, Will, and Ariel. *Our Oriental Heritage*. NY: Simon & Schuster, 1935 - 1975.

E   Eisler, Colin. *David's Songs - His Psalms and Their Stories*. NY: Dial Books, 1992.

J   Glubok, Shirley. *The Art of the Lands in the Bible*. NY: Atheneum, 1963.

HF   Ingram, Tolbert R. *Maid of Israel*. Nashville, TN: Broadman Press, 1955.
Y   Story of Naaman and the Israelite slave girl.

HF   Kubie, Nora Benjamin. *King Solomon's Horses*. NY: Harper, 1956.

HF   Kubie, Nora Benjamin. *King Solomon's Navy*. NY: Harper & Brothers, 1954.
J   Young Jared earns his freedom serving in the king's Navy.

HF   Levitin, Sonia. *Escape from Egypt*. Boston: Little, Brown, 1994.
Story of the Exodus

E   Lorenz, Graham. *A Road Down in the Sea*. NY: Thomas Y. Crowell Co., 1946.

HF/J   Malvern, Gladys. *The Foreigner, the story of a Girl named Ruth*. NY: Longmans, Green and Co., 1954.

HF/J   Malvern, Gladys. *Saul's Daughter*. NY: Longmans, Green and Co., 1956.

R   National Geographic Book services, Ed. *Everyday Life in Bible Times*. Washington: National Geographic Society, 1977.

J   Odijk, Pamela. *The Israelites*. Englewood Cliffs, NJ: Silver Burdett Press, 1990.

J   Patterson, Jose. *Angels, Prophets, Rabbis, and Kings - From the Stories of the Jewish People*. NY: Peter Bedrick Books, 1991.

R   Pfeiffer, Charles F. *Old Testament History*. Grand Rapids, MI: Baker Book House, 1973.

J   Pritchard, J.B. *The Ancient Near East in Pictures relating to the Old Testament*. Princeton: Princeton University Press, 1955.

HF   Rivers, Francine. *Unashamed*. Wheaton, IL: Tyndale House, 2000.
A   Story of Rahab who hid the spies in Jericho
        Lineage of Grace series

HF   Rivers, Francine. *Unshaken*. Wheaton, IL: Tyndale House, 2001.
        Story of Ruth
        Lineage of Grace series

E   Sanderson, Ruth. *Tapestries - Stories of Women in the Bible*. NY: Little, Brown and Co., 1998.

R   Tenny, Merrill C. *The Zondervan Pictorial Bible Dictionary*. Grand Rapids: Zondervan, 1967.

R   Winer, Bart. *Life in the Ancient World*. Random, 1961.

V   Samson and Delilah. Videocassette. Dir. DeMille. Paramount, 1951. 128 min.

V   The Ten Commandments. Videocassette. Dir. DeMille. Paramount, 1956. 219 min.

V   Moses. Videocassette. Dir. Roger Young. Turner Home Entertainment, 1996. 188 min.

Unit 14 Ancient Greeks

HF   Alcock, Vivien. *Singer to the Sea God*. NY: Delacorte Press, 1992.
J   Phaidon's life is intertwined with the gods and monsters of Greek mythology.

J   Caselli, Giovanni. *The First Civilizations*. NY: Peter Bedrick Books, Harper & Row, 1985.

J   Colum, Padraic. *The Children's Homer - The Adventures of Odysseus and the Tale of Troy*. NY: Aladdin Paperbacks, 1881 - 1972.

J   Crosher, Judith. *The Greeks*. London: Macdonald Educational; Morristown, NJ: Silver Burdett, 1974.

R   Durant, Will. *The Life of Greece*. NY: Simon & Schuster, 1935 - 1975.

R   Finley, M.I. *Early Greece - The Bronze and Archaic Ages.* NY: Norton, 1981.

J   Gay, Kathlyn. *Science of Ancient Greece.* NY: Franklin Watts, 1998.

R   Homer. *The Odyssey.* Trans. Butcher and Lang, NY: Macmillan Co., 1930.

J   Powell, Anton. *Ancient Greece.* NY: Facts on File, 1989.

E   Stewig, John Warren. *King Midas.* NY: Holiday House Book, 1999.

R   Richter, Gisela. *A Handbook of Greek Art.* NY: DaCapo Press, 1987.

Unit 15 Assyrians

R   Book of "Jonah" in the Bible

R   Cottrell, Leonard. *Land of the Two Rivers.* Cleveland: World Pub. Co., 1962.

E   Haiz, Danah. *Jonah's Journey.* Minneapolis: Lerner Pub. Co., 1973.

J   Jameson, Cynthia. *The Secret of the Royal Mound.* NY: Coward, McCann and Geoghegan, 1980.

R   Saggs, H.W.F. *Everyday Life in Babylonia and Assyria.* NY: Dorset Press, 1987.

R   Saggs, H.W.F. *The Might that was Assyria.* NY: St. Martin's Press, 1990.

R   Silverberg, Robert. *The Man Who Found Nineveh; the Story of Austen Henry Layard.* NY: Holt, Rinehart and Winston, 1964.

Unit 16 Etruscans

J   Connolly, Peter. *Hannibal and the Enemies of Rome.* Morristown, NJ: Silver Burdett Co., 1978.

J   Glubok, Shirley. *The Art of the Etruscans.* NY: Harper & Row, 1967.

P   Gore, Rick, "The Eternal Etruscans", National Geographic, Vol. 173, No. 6, 6/1988.

J   Honness, Elizabeth. *The Etruscans - An Unsolved Mystery.* Philadelphia: Lippincott, 1972.

J   Leacroft, Helen, and Richard. *The Buildings of Ancient Rome.* NY: W.R. Scott, 1969.

R   Strong, Donald. *The Early Etruscans.* NY: Putnam, 1968.

Unit 17   Neo-Babylonian or Chaldean

R   Book of "Daniel" in the Bible

R   Cottrell, Leonard. *Land of the Two Rivers*. Cleveland: World Pub. Co., 1962.

R   Falls, Charles Buckles. *The First 3000 Years*. NY: Viking Press, 1962.

R   Rice, Edward. *Babylon, Next to Nineveh - Where the World Began*. NY: Four Winds Press, 1979.

R   Saggs, H.W.F. *Everyday Life in Babylonia and Assyria*. NY: Dorset Press, 1987.

P   Speiser, E. A., "Ancient Mesopotamia: A light that did not fail", National Geographic, 1/1951.

HF/A   Weinreb, Nathaniel Norsen. *The Babylonians*. NY: Doubleday & Co. Inc., 1953.
         Beladar, as physician, diplomat, and confidant of Nebuchadnezzar has many adventures and intrigues.

Unit 18   Scythians

P   Edwards, Mike, "Searching for the Scythians", National Geographic, 9/1996.

R   Stanton, Mary, and Hyma, Albert. *Streams of Civilization - Volume One*. IL: Christian Liberty Press, 1978.

Special Project   Wonders of the Ancient World

R   DeAngelis, Therese. *Wonders of the Ancient World*. Philadelphia: Chelsea House, 1999.

R   National Geographic, Ed. *Wonders of the Ancient World - Atlas of Archaeology*. Washington DC: The Society, 1994.

J   Silverberg, Robert. *The Seven Wonders of the Ancient World*. NY: Crowell-Collier Press, 1970.

R   Stewart, Desmond. *The Pyramids and Sphinx*. Newsweek, NY: Newsweek Book Division, 1971.

V   Pyramid. Videocassette. Produced by Unicorn Projects, Inc. for PBS. Dorset Video. 1988. 60 min.

V   The Seven Wonders of the Ancient World. Videocassette. Questar Video. 1989. 60 min.

TRISMS©

Unit Plans

ART	MUSIC	ARCHITECTURE
		Ark
		Questionnaire
		Whose design was being followed?
		Who was the "architect?"
.		
Cave paintings		
Questionnaire		
	Questionnaire	Using graph paper draw the
	Jubal	dimensions of the ark as
		described in Gen. 6:15
		with 1 blk=1 cubit
		Çatal Hüyük
		Answer parts II & III on
		Questionnaire
HISTORICAL EVENTS		
Creation		
Expulsion from Eden		
The Flood		
Tower of Babel		
Dispersion		
Wheat and barley grown in Egypt		
Rice cultivation in China		
Corn cultivation in Mexico		

SCIENCE	CIVILIZATIONS	LITERATURE	VOCABULARY
S-1a		L-1a	prehistory
Archaeology		Tape 1 IEW	millennium
		Note making and outlines	archaeology
			anthropology
		L-1aa	artifact
		Book selection	chronology
			excavate
			migrate
S-1b		L-1b	dispensation
Archaeology		Creation stories	postdiluvian
		Key word outline	
			circa
			zenith
			culture
			civilization
			mnemonics
	Worksheet 1	L-1c	decipher
		Summarizing from notes	scaffold
			debris
			agrarian
			lyre
			dulcimer
	Map the migration of Noah's	L-1d	
	sons Gen 10: 1-32	Analysis	
		Compare & Contrast	
	Map #2		
	Label Mt. Ararat		
		L-1e	
		Compare & Contrast	
		Banned Word List	
		L-1ee	
		Movie with Q & A	

MAP DETAILS	OTHER AREAS OF INTEREST	COMPARE QUESTION	
Tigris River	Shem the ancestor of Semites	Civilization vs Culture	
Euphrates River	Japheth the ancestor of Hittites		
Caspian Sea	Ham the ancestor of Africans		
Red Sea	Japheth-father of IndoEuropeans		
Black Sea	cultivation of grain		
Mediterranean Sea	early cities - Jericho (Palestine)		
Persian Gulf	domestication of animals		
Indian Ocean	stone age people		

	ART	MUSIC	ARCHITECTURE
	Jewelry from the Royal tombs		Ziggurat
	Questionnaire		Questionnaire
			What did the ziggurat
			symbolize?
		Questionnaire	Mosaics of Ur
			Mosaics were created in the
			sides of ziggurats with clay
			cones. These were a
			regular feature of the
			Sumerian ziggurat.
			What did this mosaic look
			like? What did their colors
			symbolize? How did these
			ziggurats affect the
			appearance of their city?
			Mahenjo-Daro
			elaborate sanitation systems
			Answer II & III on the
			Questionnaire
	Stealite seals		
	Questionnaire		
	HISTORICAL EVENTS		
	4000 Sumerian pictograph		
	4000 King Urakagina repeals taxes and tax collectors of Lagash		
	3500 Sumerian cuneiform		
	3500 founding of Ur		
	3500 fishing villages flourish in Peru		
	3500 Haida culture begins in NW coast of Canada		
	3000 wheel appears		
	2700 King Gilgamesh reigns Uruk in Sumer		

SCIENCE	CIVILIZATIONS	LITERATURE	VOCABULARY
	4000-1750 BC	L-2a	Mesopotamia
	Sumerians	Openers	Semitic
	Questionnaire	Tape 3 IEW	fertile crescent
			city-state
	map extent of empire	L-2aa	cuneiform
	Map #2 Label Ur	Cuneiform	epic
	Add Map Details	Expository essay	glaze
			alloy
S-2a	What discovery allowed	L-2b	smelt
Smelting	this nomadic people to	Narrative Poetry	textile
	become agrarian?	Epic of Gilgamesh	alluvium
			ore
			ziggurat
	Worksheet 2		mosaic
			lyre
S-2b	3000-1500 BC	L-2c	dominate
Cotton	Indus Valley (River people)	Analysis	invasion
	short form		federation
			nomad
	map extent of empire		seal
	Map #3		trademark
	Label Mahenjo-Daro		citadel
			steppes
S-2c	Label map #3	L-2d	domesticate
Bonus	Indian Ocean	Q & A	science
	Bay of Bengal		technology
	Arabian Sea	L-2dd Bonus	destiny
	Himalaya	Movie poster	fate
	Ganges River		
	Khyber Pass		
	Mark map #1	L-2e	
	mark all co-existing	Book report	
	civilizations		
		L-2ee	
		Movie with Q & A	

MAP DETAILS	OTHER AREAS OF INTEREST	COMPARE QUESTION	
Red Sea	Math-geometry	Compare a ziggurat to	
Mediterranean Sea	sexagesimal system	the Tower of Babel.	
Black Sea	Scribe - Amat-Mamu		
Caspian Sea	Enheduana		
Persian Gulf	Akkadians		
Tigris River	Queen Shudi-Ad		
Euphrates River	Urnanshe		
Indian Ocean	Kubaba		

	ART	MUSIC	ARCHITECTURE
			Step Pyramid
			Questionnaire
			Architect Imhotep
			Optional:
			Parts II & III on an Egyptian city
	Tomb Art	Questionnaire	
	Questionnaire	combine Ancient & Old Kingdoms	
	Draw a 2-D picture & paint		Sphinx or
	with solid colors (tempera)		Great Pyramids
	No mixing or shading		Questionnaire
			Optional:
			Parts II & III on an Egyptian
			village
	HISTORICAL EVENTS		
	4000 Sails first used on boats on the Nile River		
	3500 Naqada culture begins in Egypt		
	3000 Scientist able to predict the size of the annual flood		
	2500 libraries in Egypt		
	2800 Stone Age in Britain		
	2650 - 2500 great period of pyramid building		
	2200 Middle Joman period begins in Japan		

TRISMS

SCIENCE	CIVILIZATIONS	LITERATURE	VOCABULARY
S-3a	4000-2686 BC	L-3a	dynasty
The Nile River	Ancient Egypt	Hieroglyphics	scribe
	pre-dynastic period		delta
		L-3aa	pictograph
	map extent of empire	Book selection	papyrus
	Map #4 Label Memphis		relief (n)
	Add Map Details		two-dimensional
			empire
S-3b	Combine the Ancient & Old	L-3b	cataract
Egyptian calendar	Kingdoms together on one	Poetry	causeway
	Questionnaire		
			audit
			shadoof
			ideograph
			nepotism
			sculpture
S-3c	2686-2155 BC	L-3c	quarry
Tools	Old Kingdom	Dress-ups	obelisk
	overthrow of 6th dynasty	Tape 2 IEW	adze
			plumbline
	map extent of empire		harp
	Map #4		
	Label the Great Pyramids		
	Worksheet 3	L-3d	
		Tribute	
		2600-2300 BC	
		collections of Egyptian	
		religious literature carved	
		in pyramids	

MAP DETAILS	OTHER AREAS OF INTEREST	COMPARE QUESTION	
Nile River	The Nile River	Compare the Step	
Red Sea	Architect -- Imhotep	Pyramid with the Great	
Mediterranean Sea	Annual audit for taxation	Pyramids.	
Nubian Desert	Queen Nitocris		
Sahara Desert	Why were their deities part		
1 - 4 Cataracts	human and part animal?		
Blue Nile River	Queen Meryet-Neith		
White Nile River	Dr. Merit Ptah		

	ART	MUSIC	ARCHITECTURE
			Temple of Amon at Luxor
			Questionnaire
			Optional:
			Parts II & III on an Egyptian City
	Colossal statues		Optional: Draw examples of the
	Questionnaire		various types of
			Egyptian columns
			Papyrus, Bell, Palm, Lotus
		Questionnaire	
		combine Middle and New	
		Kingdoms	
			Temple of Hatsheput at
			Deir el Bahri
			Questionnaire
			Architect Senmut
			Optional:
			Parts II & III on a
			Nobleman's estate
	Limestone bust of Neferetiti		
	Questionnaire		

HISTORICAL EVENTS		
2000 End of Sumerian power in Mesopotamia		
2000 Bronze Age in Scotland		
1935 Abraham visits Egypt		
1786 Rule of the Hyksos in Egypt		
1500 Shang kingdom in China		
Akhenaten's endeavors consume the treasury.		
He loses the tribute of Syria & Palestine.		
770 Nubian rulers in Egypt		

SCIENCE	CIVILIZATIONS	LITERATURE	VOCABULARY
	2040-1786 BC	L-4a	irrigation
	Middle Kingdom	Egyptian novel	inundation
		Analysis	immortality
	map extent of empire		cartouche
	Map #4	L-4aa Bonus	incense
	Label Thebes & Luxor	Movie	myrrh
	Add Map Details	Story elements	ebony
			frieze
S-4a	Combine Middle Kingdom	1580 - 1350 BC	survey
Irrigation	with New Kingdom on	Book of Dead existed	colossi
	same questionnaire	in pyramid inscriptions	plunder
			predestination
		L-4b	mummy
		Decorations	
		Tape 4 IEW	
S-4b	1567-644 BC	L-4c	
Embalming	New Kingdom	Creative Writing	
	map extent of empire		
	Map #4		
	Label the Valley of the Kings		
S-4c	Worksheet 4	L-4d	
Bonus		Magazine interview	
	Mark Map #1	L-4e	
	mark all co-existing	Book report	
	civilizations	Tape 6 IEW	

MAP DETAILS	OTHER AREAS OF INTEREST	COMPARE QUESTION	
Red Sea	Joseph	Compare Queen	
Nile River	Moses	Hatsheput's obelisk with	
Mediterranean Sea	Queen Ety of Punt	that of male pharaohs.	
Nubian Desert	Queen Tye		
Sahara Desert	Queen Ahotep		
1 - 4 Cataracts	beautician - Henut		
Blue Nile River	Hetepheres		
White Nile River	Khentkawes		

Unit 5

	ART	MUSIC	ARCHITECTURE
			Solomon's Temple
			palm shaped columns
			Questionnaire
	Glass vase		
	Questionnaire		
		Questionnaire	

HISTORICAL EVENTS		
2700 Founding of Troy		
2000 Warlike Mycenaeans		
1400 Phoenician alphabet		
1300 Hebrews enter Canaan		
860 Phoenicians conquered by Babylon		
850 Carthage founded by Elissa		
Phoenicians develop ram and bireme		

TRISMS©

SCIENCE	CIVILIZATIONS	LITERATURE	VOCABULARY
	3000 - 860 BC	L-5a	cartography
	Phoenicians (Sidonians)	Phoenician script	custom
	Questionnaire		caravan
			colony
	map extent of empire		merchant
	Map #5		property (science)
	Label Tyre, Sidon, & Byblos		pirate
			indemnity
S-5a		L-15b	diadem
Glass		Book selection	mercantile
			script
			topheth
			meghazil
			flax
S-5b	850 - 146 BC	L-5c	
Glass	Carthage (W. Phoenicia)	Monologue	
	short form		
	Add Carthage to map #5		
S-5c	Worksheet 5	L-5d Bonus	
Bonus		Biblical references	

MAP DETAILS	OTHER AREAS OF INTEREST	COMPARE QUESTION	
Mediterranean Sea	Sea Trade routes	In what way did the	
Atlantic Ocean	Navigation	Phoenician's script link the	
Pillars of Hercules	Jezebel	known world? How was	
Iberian Peninsula	Hebrews	it different from hieroglyphics	
Balkan Peninsula	Mercantile system	and cuneiform?	
Apennines Peninsula	Hannibal		
Anatolia	Punic Wars		
Atlas Mountains	Sophonisba		

TRISMS©

	ART	MUSIC	ARCHITECTURE
			Palace of Knossos
			Questionnaire
	Polychrome Kamares ware		Bonus: What sorts of
	with the marine motif		elaborate systems did this
	Questionnaire		great palace have?
		Questionnaire	
	HISTORICAL EVENTS		
	3000 Bronze Age begins on Crete		
	2200 First navy		
	2000 Pottery wheel introduced		
	2000 Warlike Mycenaeans		
	1450 Mycenaean invasions		
	1450 Volcano eruption on Thera		
	1375 Knossos burnt to the ground		

SCIENCE	CIVILIZATIONS	LITERATURE	VOCABULARY
	3500 - 1400 B.C.	L-6a	fresco
	Minoans (Cretans)	Minoan script	linear
	Questionnaire		labyrinth
			legend
	map extent of empire		cataclysm
	Map #6		motif
	Label Knossos		pumice
			magma
S-6a		L-6b	tsunami
Volcanoes		Myths	coldera
			tephra
			polychrome
			veneer
			invert
S-6b	Worksheet 6	L-6c	
Simple Machines		Myths	
	How are the columns of the		
	Minoan architecture		
	different from the		
	Egyptian columns?		
	Diagram each style.		
		L-6d	
		News story	
	Mark map #1	L-6e	
	mark all co-existing	Book report	
	civilizations		

MAP DETAILS	OTHER AREAS OF INTEREST	COMPARE QUESTION	
Rhodes	Mycenaeans	Compare the fresco styles	
Crete	Sea Trade routes	of Egypt and Crete.	
Peloponnesus Peninsula	Labyrinth		
Aegean Sea	Sir Arthur Evans		
Mediterranean Sea	Eteocretans		
Black Sea	Kydonians		
Ionian Sea	Pelasgians		
	Daedalus		

TRISMS©

	ART	MUSIC	ARCHITECTURE
			Fill out only the second half of questionnaire regarding housing
			Parts II and III
	Ritual art / bronze		
	Questionnaire		
		Questionnaire	
		combine Shang & Chou	
	Lacquer work		
	Questionnaire		
	HISTORICAL EVENTS		
	Emperor Huang-ti		
	2700 Empress Hsi-Ling discovers silk		
	2000 astronomical observatory		
	1200 Destruction of Troy by the Greeks		
	1100 - 900 First Assyrian civilization in northern Mesopotamia		
	1000 Banki culture in Japan		
	1000 Dong Son culture in SE Asia		
	551 Birth of Confucius		

SCIENCE	CIVILIZATIONS	LITERATURE	VOCABULARY
	1500 - 1027 BC	L-7a	laissez-faire
	Shang	Chinese folklore	barter
	short form		acupuncture
		L-7aa	animism
	map extent of empire	Book selection	apothecary
	Map #7		moxibustion
			divination
			oracle
S-7a		L-7b	feudal system
Silk		Summarizing Narrative	gnomon
		Stories	
		Tape 2 IEW	abacus
			thatch
			alchemy
			crucible
			logograph
S-7b	1027 - 257 BC	L-7c	zither
Acupuncture	Chou or Zhou	Creative writing	plectrum
	short form	Tape 5 IEW	lacquer
			Taoism
	map extent of empire		Confucianism
	Map #7		
	Worksheet 7	L-7d Bonus	
		Comparison	
		L-7e	
		Legend	

MAP DETAILS	OTHER AREAS OF INTEREST	COMPARE QUESTION	
Pacific Ocean	Calendars predicting eclipse	Why did this civilization not	
Yellow River	Legendary times	show much change until the	
Yangtze River	Mythic culture	20th c. AD?	
Kunlun Mountains	Oracle bones	What has not changed to this	
Mekong River	Bronze age	day?	
Taklimakan Desert	Fu-Hi		
Gobi Desert	Shen-nung		
	feudal system		

ART	MUSIC	ARCHITECTURE
Hittite Iron jewelry		
Questionnaire		
		King's Gate at Boghzkaÿ
		Questionnaire
		Part I & II on Hittites
	Questionnaire - Aryan	
HISTORICAL EVENTS		
Labarnas the Hittite founder		
Egyptian Pharaoh Thotmes III forced to pay tribute to Hittites		
1530 Hittite Empire reaches Babylon		
1300 Vedic hymns composed		
1190 Invasion of the sea people		
1120 City of Mycenae destroyed		
late 800's Phrygians inherit Hittite civilization		
Aryans subdue Semities, Amorites, and Bedouins		

TRISMS©

SCIENCE	CIVILIZATIONS	LITERATURE	VOCABULARY
S-8a	2000 - 1150 BC	Prayers in times of	monopoly
Iron Age	Hittites	Plague by Mursilis II	successor
	short form		commonwealth
			constitution
	map extent of empire	Hatusilis wrote an	diplomacy
	Map #2 Label Boghzkaÿ	autobiography.	facade
	Add Map Details		metropolis
			basalt
S-8b	Worksheet 8	L-8a	high relief
Iron		Triple extensions	ductile
		Tape 4 IEW	
	2000-600 BC	L-8b	oblation
	Aryan (India)	Hymn	Hinduism
	Questionnaire		caste
			Indo-European
	map extent of empire		invocation
	Mark map #3		metallurgy
			patron
	Mark map #3	L-8c	
	Indian Ocean	Comparisons	
	Bay of Bengal		
	Arabian Sea		
	Ganges River		
	Himalaya		
	Kyber Pass		
	Mark map #1	L-8d Bonus	
	mark all co-existing	Expository essay	
	civilizations		
		L-8e	
		Book report	

MAP DETAILS	OTHER AREAS OF INTEREST	COMPARE QUESTION	
Mediterranean Sea	Esau married a Hittite woman.	Compare the Hittite common-	
Red Sea	Queen Puduhepa	wealth to an empire.	
Black Sea	Mastigga		
Caspian Sea	Phrygians King Midas		
Persian Gulf	Haburi warriors		
Tigris River	First World-historical War		
Euphrates River	Tudhaliyas II develops chariot		
Indian Ocean	Telipinus the law giver		

Unit 9
First Test and Special Project

	ART	MUSIC	ARCHITECTURE
	Stele of Hammurabi		
	Questionnaire		
			Bridge over the Euphrates
			Questionnaire
			How wide was the
			Euphrates River?
	HISTORICAL EVENTS		
	Land of Shinar		
	2000 Inuits in Arctic		
	2000 Andean settlements in Peru		
	2000 Hittites settle in Anatolia		
	1595 Hittites plunder Babylon		
	1500 Collapse of Minoan civilization on Crete		

SCIENCE	CIVILIZATIONS	LITERATURE	VOCABULARY
no science	2000-1600 BC	L-9a	code
	Old Babylonian Empire	Poetry	law
	Questionnaire		stele
		L-9aa	caisson
	map extent of empire	Book selection	omen
	Map #2		zodiac
	Label Babylon		astronomy
			inscribe
	Hammurabi 6th King of	L-9b	incantation
	Babylon	Code of Hammurabi	
	Worksheet 9		
	Select a topic from the	L-9c Bonus	
	Topics list.	Compare	
	These are the mysteries of		
	the ancient world.		
	Locate & Label these on Map #1		
	This project can be a report,		
	a model, or diagrammed		
	poster.		
	If you choose a report,		
	present the experts' theories		
	as well as your own		
	thoughts on the subject.		

MAP DETAILS	OTHER AREAS OF INTEREST	COMPARE QUESTION	TOPICS
Red Sea	Akkadians	Compare their justice system	Atlantis
Mediterranean Sea	Sumeria	with the Sumerians.	Stonehenge
Anatolia	Hittites		Easter Island
Caspian Sea	Kassites		Chalk cliff drawings
Persian Gulf	Assyria		Nazca lines
Tigris River	Grand Canal		Bighorn medicine wheel
Euphrates River	Doctors & Surgeons		Nan Modal
Arabian Peninsula	Ennigaldi		

	ART	MUSIC	ARCHITECTURE
			Citadel of Mycenae
			Questionnaire
		Questionnaire	
	Gold Death Mask		
	Questionnaire		

HISTORICAL EVENTS		
1400 Mycenaeans become leaders in the Aegean		
Bronze work appears in central Europe		
Bantu in West Africa		
1379 New Kingdom in Egypt		
1375 Knossos destroyed ending Minoan civilization		
1230 Troy destroyed		
1200 Mycenaeans overcome by the Dorians		

Unit 10

SCIENCE	CIVILIZATIONS	LITERATURE	VOCABULARY
	1600-1120 BC	L-10a	fiefdom
	Mycenaean	The Iliad	centralize
	Questionnaire	Analysis & Summary	Achaian
			acropolis
	map extent of empire		elucidate
	Map #6		granulation
	Label Mycenae & Troy		toreutic
			cyclops
S-10a	Worksheet 10	L-10b	plaster
Gold		Character analysis	lintel
		L-10c	
		Iliad Q & A	
		L-10d	
		The Iliad	
		Historical significance	
		L-10e	
		Book report	
		L-10ee bonus	
		Genealogy	

MAP DETAILS	OTHER AREAS OF INTEREST	COMPARE QUESTION	
Rhodes	Trojan War	There is a controversy among	
Crete	Dorians	scholars. Some feel the	
Peloponnesus Peninsula	Funeral Games	Minoans are ancestors to the	
Balkan Peninsula	Cyclopean masonry	Mycenaeans. Some do not.	
Aegean Sea	Beehive Tombs	What facts support each view?	
Mediterranean Sea	Heroic Age		
Black Sea			
Ionian Sea			

TRISMS©

ART	MUSIC	ARCHITECTURE
Olmec Floor mosaics	Questionnaire - Olmec	
Questionnaire		
Olmec Mega sculpture		Circular Pyramid
Basalt heads		Questionnaire
Questionnaire		
		Great Pyramid at LaVenta
Nubian Kerma Ware Pottery		
Questionnaire		

HISTORICAL EVENTS		
1600 BC Kerma age of Kush		
1000 LaVenta becomes an important Olmec center		
725 Kush conquers Egypt		
667 Assyrians defeat Kush		
530 Meroë the new capital		
270 BC - A D 350 Meroë age		
543 - 580 Christian Kush		
1960 Lake covers Kush		

SCIENCE	CIVILIZATIONS	LITERATURE	VOCABULARY
S-11a	1200 - 500 BC	L-11a	primitive
Agriculture	Olmec	Script	Mesoamerica
	1st complex society in		heritage
	North America	L-11aa	ecology
	short form	Book selection	cultivate
			sanctuary
			serpentine
			domesticate
S-11b	map extent of the empire	L-11b	ritual
Agriculture	Map #8	Compare & Contrast	deface
	Label LaVenta	Descriptive essay	
	Add Map Details	Tape 4 IEW	
	Worksheet 11		
S-11c	2000 - 200 BC	L-11c	kiosk
Agriculture	Nubia (Kush/Cush)	Nubian Proverb	wadi
	short form		homage
			vanguard
	map extent of empire		prostrate
	Map #4		pinnacle
	Label Kerma & Meroë		potentate
	Label Map #4	L-11d	
	Red Sea	Script writing	
	Nile River		
	1 - 6 Cataracts		
	White Nile River		
	Blue Nile River		
	Nubian Desert		
	Mark map #1		
	mark all co-existing		
	civilizations		

MAP DETAILS	OTHER AREAS OF INTEREST	COMPARE QUESTION	
Pacific Ocean	Theories of origin - Olmec	Compare the Olmec temples	
Gulf of Mexico	Hyksos	with Sumerian ziggurats.	
Yucatan Peninsula	Axum		
Caribbean Sea	Taharka-defender of Israel		
Bay of Campeche	Queen Amanirenas		
	Sudan		
	Symbiotic civilizations		

	ART	MUSIC	ARCHITECTURE
			Tent Tabernacle
			Questionnaire
		Questionnaire	
		Exodus 15	
	Ark of the Covenant		
	Questionnaire		
	HISTORICAL EVENTS		
	Pharaoh enslaves Israel through delinquency and/or tax rebellion		
	1250 Exodus		
	Moses receives the Ten Commandments		
	1150 Gideon		
	1085 Samson		
	End of New Kingdom in Egypt		
	1085 - 945 Government of Egypt passes to pharaohs in the north		
	1100 - 900 Assyrian civilization declines in Iraq		

SCIENCE	CIVILIZATIONS	LITERATURE	VOCABULARY
	Exodus - 1025 BC	L-12a	covenant
	Hebrews/Israel	Short story	famine
	Nomadic period		patriarch
	Questionnaire		theocracy
		L-12aa	monotheism
		Movie Review	Pentateuch
			Torah
			sojourner
	Map their wanderings after	L-12b	append
	the Exodus from Egypt	Evaluation	exodus
	Map #9		apostasy
			judge
			idol
	Mark where the various	L-12c	
	tribes settled on map #9	Contrast	
S-12a	Worksheet 12	L-12d	
Nutrition		Story outline &	
		Character sketches due	
S-12b		L-12e	
Nutrition		Book report	

MAP DETAILS	OTHER AREAS OF INTEREST	COMPARE QUESTION	
Egypt	Land of Canaan	What made the Hebrews	
Red Sea	Judges of Israel	different from every other	
Mount Sinai	Sabeans	civilization? Look at their	
Gulf of Aqaba	Persia/Medes	religion, lifestyle, laws,	
Gulf of Suez	I Samuel 20:17	strategies, and conquest.	
Jordan River	Egypt	Compare the Ten Command-	
Negev Desert	Arabia	ments with the Code of	
Dead Sea	Amorites	Hammurabi.	

TRISMS©

	ART	MUSIC	ARCHITECTURE
			Solomon's Palace
			Questionnaire
		Questionnaire	Besides the Temple and Palace, why were there no significant contributions to architecture?
		I Samuel 18:6 & 7	
		I Chronicles 15: 16-22	
		II Chronicles 5:11-14	
HISTORICAL EVENTS			
1025 Saul			
1000 - 968 David			
968 - 922 Solomon			
922 Kingdom divided			
Rehoboam rejected because of taxation			
853 Battle of Qarqaar Assyria defeated by Israel			
722 Assyria conquers northern kingdom of Israel			
586 Judah destroyed by Babylonians and taken captive			

SCIENCE	CIVILIZATIONS	LITERATURE	VOCABULARY
no science	1025 - 926 BC	L-13a	prophecy
	Israel (the Kingdom)	Psalms	precept
	Questionnaire		prophet
		L-13aa	Diaspora
		Book selection	piety
			virtue
			rectitude
			psalm
	Mark the divided kingdom	L-13b Bonus	proverb
	of Israel	Biblical references	sanctuary
	Map #9		parable
		L-13bb	Messiah
		Story rough draft due	
	Worksheet 13	L-13c	
		Paraphrasing &	
		Descriptive summary	
		587 Early books of Old	
		Testament assembled	
		during Babylonian	
		captivity	
		L-13d	
		Editing and titles	
		L-13dd	
		Type short story	
	Mark map #1	L-13e	
	mark all co-existing	Book cover	
	civilizations		

MAP DETAILS	AREAS OF INTEREST	COMPARE QUESTION	
Mediterranean Sea	Philistines	Compare the tent tabernacle	
Mount Sinai	J	with the Temple built by	
Dead Sea	Queen Makeda of Sheba	Solomon.	
Sea of Galilee	frieze of Sennacherib		
Red Sea	Diaspora		
Jerusalem	wealth through taxation		
Bethel	Rehoboam's rejection		
Damascus			

TRISMS©

ART	MUSIC	ARCHITECTURE
Geometric style as seen in pottery		Nothing has survived
Questionnaire		
Draw examples of the basic shapes of Greek vases.		
List the names of the basic shapes.		
	Questionnaire	
Black-figure pottery		The Temple of Artemis at
Questionnaire		Eleusis 600 BC or earlier
		Questionnaire
Red-figure pottery		
Questionnaire		
HISTORICAL EVENTS		
1100 Dorians invade Greece		
1000 - 700 Geometric Age		
776 1st Olympics		
800 Etruscans set up city-states		
735 Syracuse founded in Sicily		
650 Cyrene founded in North Africa		
590 Solon introduces laws abolishing enslavement for debtors		
508 Cleisthenes introduces democratic reforms to Greece		

SCIENCE	CIVILIZATIONS	LITERATURE	VOCABULARY
no science	1150-500 BC	L-14a	votary
	Ancient Greece	Homer	commemorate
			terra cotta
			emboss
	1150 - 750 BC		engrave
	Greek Dark Ages		replica
	short form		burnish
			foreshorten
	Map the three principle	L-14b	exploits
	people groups of this region	Character analysis	suitor
	and time period.		
	Map #6		
	c. 650-500 BC	750 Hesiod's Theogony	symmetric
	Archaic Greece	collection of stories	oligarchy
	short form	about the creation of the	isthmus
		world and gods of	peninsula
		Mythology	agora
			archipelago
		L-14c	ostracize
		Odyssey Q & A	panpipe
	Worksheet 14	L-14d	hoist
		Opinion	
		700 Homeric poems	
		Iliad & Odyssey	
		L-14e	
		Book report	

MAP DETAILS	OTHER AREAS OF INTEREST	COMPARE QUESTION	
Rhodes	Herakles	Why is this period called the	
Crete	Sappho	Greek "Dark Ages"?	
Peloponnesus Peninsula	Pesistratus the Tyrant	What characterizes a	
Balkan Peninsula	Solon of Athens	dark age compared to an	
Aegean Sea	Aeschylus	enlightened age?	
Mediterranean Sea	Homer		
Black Sea	divine genealogies		
Ionian Sea	Pythagoras		

Unit 15

	ART	MUSIC	ARCHITECTURE
			Palace at Nineveh
			"The Incomparable" or
			Sargon's Palace at Nimrud
			Questionnaire
		Questionnaire	
	Bas-relief of battle scenes in the Palace at Nineveh shows animals in motion		
	Questionnaire		
	HISTORICAL EVENTS		
	750 Israel rebels against paying tribute and is subdued and fined		
	734 Israel rebels against paying tribute and is taken captive		
	724 - 705 Sargon rules Assyria		
	704 - 681 Sennacherib rules Assyria		
	671 Egypt overrun by Assyria		
	612 Nineveh sacked by the Babylonians and Medes		
	609 Assyria defeated by Babylon		

SCIENCE	CIVILIZATIONS	LITERATURE	VOCABULARY
	1500-612 BC	L-15a	deport
	Assyria	Oratory	dominion
	Questionnaire		edifice
			terrorism
			imperialism
	map extent of empire		reprieve
	Map #2 Label Nineveh		bas-relief
			lute
	Worksheet 15	L-15b	repent
		Feature story	allegory
			alabaster
S-15a		L-15c	
Plant classification		Writing with pictures	
		Tape 4 IEW	
S-15b			
Bonus			
	Mark map #1		
	mark all co-existing		
	civilizations		

MAP DETAILS	OTHER AREAS OF INTEREST	COMPARE QUESTION	
Tigris River	psychological warfare	Briefly discuss the	
Euphrates River	siege engines	development of the	
Persian Gulf	unification of the Near East	Mesopotamia region from	
Arabian Peninsula	Kurds	Sumeria to Assyria.	
Anatolia	Mitanni		
Cyprus	Queen Ashursharrat		
Red Sea	Queen Na'qia		
Mediterranean Sea	Queen Semiramis		

	ART	MUSIC	ARCHITECTURE
		Questionnaire	
	Bonus: Compare the frescoes of the Etruscans with those of the Minoans.		Mound Tombs of Eturia there are many to choose from
			Questionnaire
	Bucchero pottery		
	Questionnaire		
	HISTORICAL EVENTS		
	800 Etruscan city-states founded		
	616 - 578 Etruscan King Tarquinius Priscus rules Rome		
	500 Rome free of the Etruscans		
	396 Veii conquered by Rome		
	390 Gauls attack the North		
	Samnites take the South		
	Greeks take the islands		
	Rome takes what is left of Eturia		

SCIENCE	CIVILIZATIONS	LITERATURE	VOCABULARY
	800- 400 BC	L-16a	mural
	Etruscans	Script	filigree
	Questionnaire		tumuli
			toga
	Map #10		amber (n)
	map the extent of the empire		indigenous
	Label Rome		malaria
			sepulcher
		L-16b	mural
		Summarizing References	liturgy
		Tape 3 IEW	cryptograph
			generic
			sarcophagus
	Worksheet 16	L-16c	
		Personal letter	
		Tape 5 IEW	
S-16a			
Embalming			

MAP DETAILS	OTHER AREAS OF INTEREST	COMPARE QUESTION	
Apennines Peninsula	Tanaquil	Compare the reasons and	
Sicily	Greeks of Cumae	methods of embalming for	
Corsica	Vulca of Veii-sculptor	the Etruscan & Egyptian	
Sardinia	Maecenas	burial rituals.	
Elba	Lydia		
Adriatic Sea	Samnites		
Tyrrhenian Sea	Italic people		
Tiber River	Gauls		

	ART	MUSIC	ARCHITECTURE
		Questionnaire	
			Hanging Gardens
			Questionnaire
	Enameled tile		
	Ishtar Gate		
	Questionnaire		
	HISTORICAL EVENTS		
	625 BC Assyrians driven out of Mesopotamia		
	Nebuchadnezzar II claims Syria and Palestine for Babylon		
	605 War with Egypt		
	600 Nok people of West Africa mine iron		
	586 Fall of Jerusalem		
	557 Persian empire founded		
	539 Cyrus of Persia liberates Babylon		

SCIENCE	CIVILIZATIONS	LITERATURE	VOCABULARY
	612-539 BC	L-17a	apex
	Neo-Babylonian (Chaldeans)	Short story	lexicon
	Questionnaire		eunuch
			harem
	map extent of empire		subsidiary (n)
	Map #2		expansionist
	Label Babylon		suzerain
			liberate
		L-17b	hierarchy
		Persuasive essay	superstitious
			stereotype
			hydraulic
			asphalt
			mongrel
			neo
S-17a		L-17c	
Hydraulics		Summary	
S-17b	Worksheet 17	L-17d	
Hydraulics		Summary	
		L-17dd	
		Comparison	
S-17c	Mark map #1	L-17e Bonus	
Bonus	mark all co-existing	Debate	
	civilizations		
		L-17ee	
		Story outline &	
		Character sketches due	

MAP DETAILS	OTHER AREAS OF INTEREST	COMPARE QUESTION	
Red Sea	Hurrians	Contrast the end of the	
Mediterranean Sea	Cassites	Chaldean empire with	
Black Sea	Elamites	that of those they conquered.	
Caspian Sea	Aramaens		
Persian Gulf	Adad - Guppi		
Tigris River	Amytis		
Euphrates River	Josiah of Judah		
Indian Ocean	Johoiakim of Judah		

Unit 18
Semester Tests and Special Project

ART	MUSIC	ARCHITECTURE
Review Questionnaires	Review Questionnaires	Review Questionnaires
HISTORICAL EVENTS		
550 - 330 Persian Empire		
500 Celts in Central Europe		
326 Alexander the Great extends his empire		
285 Ptolemy rules Egypt		
265 Mauryans rule India		
250 Parthian kingdom		
221 Chinese Emperor Qin builds the Great Wall driving back the Huns		
Roman conquest		

TRISMS©

SCIENCE	CIVILIZATIONS	LITERATURE	VOCABULARY
no science	800 BC - AD 200	L-18a	barbarian
	Scythians	Origin myth	pillage
	short form		enigma
			vassal
	Map # 11		tattoo
	Add Maps Details		Amazon
			cauterize
			koumiss
	Worksheet 18	L-18b	kurgan
		Comparison	
		L-18bb	
		Short story	
		Rough draft due	
	Select a topic subject from the Topics list.	L-18c	
	These are the seven Wonders	Herodotus' account	
	of the Ancient World.		
	Map their locations on map #5		
	If you choose a report, present	L-18d	
	the experts' theories as well as	Comparison	
	your own thoughts on the		
	subject.		
	This project can be a report, a		
	model, or diagrammed poster.		
	Be sure all work in notebook	L-18e	
	is current.	Book cover	

MAP DETAILS	OTHER AREAS OF INTEREST	COMPARE QUESTION	TOPICS
Crimean Peninsula	Amazons	Compare and contrast	Mausoleum at Halicarnassus
Caucasus Mountains	Sarmations	the lifestyles of the	Lighthouse at Alexandria
Black Sea	Goths	Scythians with that of the	Colossus of Rhodes
Caspian Sea	Greek historian - Herodotus	nomadic Hebrews.	Statue of Zeus at Olympia
Aral Sea	Enaries		Temple of Artemis at Ephesus
Aegean Sea	Queen Tomyris		Great Pyramids
Danube River			Hanging Gardens of Babylon
Volga River			

Science

Assignments

S-1a: Archaeology

1. Write a complete definition of archaeology.
2. What is the difference between archaeology and anthropology?
3. What changes have occurred in the purpose and methods of the science of archaeology since the early 1900's? Who pioneered these changes?
4. What professions are represented on an archaeological team? Give a brief description of their function at a dig.
5. List some of the more famous archaeological sites that have been uncovered and the archaeologist who led the expedition.

S-1b: Discuss the importance of written records to archaeologists.

S-2a: Smelting
1. What characterizes the Bronze Age?
2. Explain the difference between pure metals and alloys.
3. Name some copper alloys?
4. How does the process of smelting work?
5. What are the elemental symbols for gold, silver, copper, and tin? What is the melting and boiling point for each?

S-2b: Cotton
The Indus Valley civilization was the first to use cotton cloth. How is cotton cultivated and manufactured?

S-2c: **Bonus:** Describe the elaborate sanitation systems of Mahenjo Daro.

S-3a: What is the source of the Nile River and where does it empty? What direction does the Nile flow?

S-3b: Describe the first Egyptian calendar; define the seasons that they developed their calendar around. Describe other methods of measuring time that the Egyptians used. Explain how they worked.

S-3c: What types of tools and surveyor's instruments were used by the Egyptian architects for building their colossal monuments?

S-4a: Irrigation
Irrigation was used in Egypt. Explain how it was used.

S-4b: The Egyptians went to great lengths to prepare their dead for burial. Explain this process and its purpose.

S-4c: **Bonus:** The Egyptians used a method of glazing for their pottery. Explain their process.

S-5a: Glass
1. What are the properties of glass?
2. What is the melting point of glass?
3. What is the difference between natural glass and synthetic glass?
Encyclopedia: glass

S-5b: Glass
1. Name the five methods used for shaping glass.
2. Explain the process of glass blowing. How is color added?
3. What methods did the Phoenicians use? Explain their process.
Encyclopedia: glass

S-5c: **Bonus:** How many different uses can you find for glass. Don't mention the obvious.

S-6a: Volcanoes
It is thought by some that the effects of a nearby volcanic eruption destroyed the Minoans.
1. What causes volcanoes?
2. What is the Ring of Fire? Map it.

S-6b: As you discovered the elaborate systems of the Minoans you realized they had an extensive understanding of simple machines. Define the basic concepts of simple machines. Make an example or draw a diagram of each.

S-7a: Silk
1. The Shang were skilled at making silk. How did they cultivate it?
2. Has this process changed since ancient times?
Encyclopedia: silk

S-7b: Acupuncture
1. Acupuncture was used during the Chou Dynasty as well as the Shang. What is acupuncture and what was it used for?
2. What were the needles made of during ancient times?
3. Is acupuncture an accepted practice among the medical community today?

S-8a: Iron Age
1. What characterizes the Iron Age?
2. Did the Hittites use pure iron or an iron alloy?
3. What is the elemental symbol for iron? What is the melting point of iron?
4. How is iron smelted today?
5. What advantages did the use of iron give the Hittites over others?
6. Go back and label each civilization you have studied as either Bronze or Iron Age. Continue for each civilization you study through Unit 18.

S-8b: Iron
1. What uses are there for pure iron?
2. What forms of iron are used today? What is the process for making each?
3. Steels are grouped into five main classifications. What are they and their uses?

S-10a: Gold
1. What are the properties of gold? What is its elemental symbol?
2. What are the uses for gold? Is it used mainly in its pure form or as an alloy?
3. Where are the main locations gold is found today?
4. How is gold mined?

S-11a: Agriculture
1. What varieties of grain and vegetables did the Olmec cultivate that were unknown outside the Americas?
2. What type of tools did the Olmec use to cultivate their fields?

S-11b: Agriculture
1. Define cultivation.
2. What is a harrow and what does it do?

S-11c: Agriculture
1. Briefly describe the process of domestication of plants.
2. What are the three main methods for developing new varieties of plants?

S-12a: Nutrition
Look up the following verses from the Old Testament and summarize what each says about nutrition: Leviticus 2:13, 3:17, 11:1. **Note:** unclean in context with scripture does not mean dirty.

Do nutritionists follow these Biblical guidelines today? Explain.

S-12b: Nutrition
1. List the four basic food groups.
2. What are nutrients? What are the five groups that nutrients are classified into?
3. Choose three foods out of each section in question one and tell what vitamins and minerals are derived from each.
4. From the vitamins and minerals listed in question two, tell specifically how each aids in the body's function.

S-15a: Plant Classification
1. The Assyrians kept a detailed record of plant life and attempted to classify them. What might have been their purpose in doing this?
2. Make a plant classification chart and a glossary defining characteristics of each group. Your first division will be Phylum, which includes vascular and nonvascular plants. The vascular plants are the divided into Subphylum. Subphylum divisions for vascular plants are Seed Plants and Nonseed Plants. The subphylum is further subdivided into Classes with seed plants subdivided as Angiosperms and Gymnosperms. Nonseed plants are subdivided into Ferns, Horsetails, and Club Mosses (sometimes called ground pine). Nonvascular plants are divided into subphyla including Mosses and Liverworts, Fungi, and Algae.

S-15b: Bonus: The Assyrians developed war machines. What were some of these?

S-16a: Embalming
1. What process did the Etruscans use for embalming?
2. What was their motive for embalming?
3. Do we embalm today?
4. In what ways do our process and motive for embalming differ from the Etruscans?

S-17a: Hydraulics

The Babylonians inherited the technical achievements of the Sumerians in irrigation and agriculture. Maintaining the system of canals, dikes, wers, and reservoirs constructed by their predecessors demanded considerable engineering knowledge and skill.

1. What type of irrigation system did the Babylonians inherit from the Sumerians? How did it work?
2. Study the geography and climate of the Mesopotamian area and explain why they needed to maintain such a developed irrigation system.

S-17b: Hydraulics

1. Define: hydraulic and hydraulics
2. Blaise Pascal (1623 - 1662), a French scientist, proposed what is now called Pascal's Law. This is the law that much of the modern science of hydraulics is based on. Write out Pascal's Law.

S-17c: Bonus: A sprinkler attached to a water hose illustrates Pascal's Law. Explain why.

Literature Assignments

An essential component of acquiring an understanding of a civilization or time period is through its literature. *Discovering the Ancient World* integrates the development of language skills using the literature of the civilizations and time period being investigated throughout the curriculum. This way, your students are not only gaining a comprehensive world view, but also learning how to analyze, interpret, and express their discoveries through writing and oral presentation.

Ψ *The Institute for Excellence in Writing* (IEW) is a video series used to teach writing structure and style. It is available directly from your TRISMS representative. We highly recommend this supplement. However, most assignments can be completed without IEW. Assignments requiring IEW are marked with a Ψ.

Skills Your Student Will Develop		
Write a tribute	Analyze a story	*Enjoy* reading
Write a book report	Write a personal letter	Look up definitions
Design a movie poster	Avoid ambiguous or overused words	Make an educated guess
Retell a story	Compare and contrast	Write a character sketch
Take notes	Do a monologue	Write a short story
Make an outline	Debate with passion	Be a movie critic
Write with style	Write a news story	Categorize literature
Use wider vocabulary	Critique a book	Recognize figures of speech
Investigate and research	Summarize a story	Paraphrase
Think critically	Use dress-ups and decorations	Compose a sermon
Identify story elements	Write essays	*Much, much more…*

Tips for Teachers
In completing the following literature assignments the student should justify all answers by examples from the literature itself. Answers should be given in complete sentences; single word answers should be discouraged. Yes/No answers should *not* be accepted. It is helpful to have the student ask "why" or "how" to get a more thorough answer. All **Bonus** assignments are optional and should be considered for extra credit.

To appreciate fully the literature of a civilization or time period it is important to have an understanding of the leading facts regarding them. For this reason we recommend the student begin the literature assignment only after having completed the Civilization questionnaire. We want students to read the literature in context with the people who wrote it. Students may need to read the selection several times before trying to answer questions. Many of the selections provided are excerpts from larger works. Students may wish to seek out and read the complete book. **All required reading is located in the Literature Selections except for Bible assignments**. For these assignments the student may use any version of the Bible.

Literature Assignments

L-1a: Ψ Taking Notes and Making Outlines
Ψ View IEW Tape 1 - Unit I
Ψ Review IEW pages 5-7: *Note Making and Outlines*
Extra practice Ψ Write a key word outline – see IEW page 6

L-1aa: Historical Fiction or Biography
📖 Select and begin reading a historical fiction or biography from the Reading Resource List. Other titles are acceptable but you should get approval from your teacher.
※ A book report will be due at the end of Unit 2. ※

L-1b: Retelling a Story
Many accounts of the stories of Creation, the Tower of Babel, and the Deluge are recorded by various civilizations.
📖 Read the stories of creation and as recounted by the Hebrews in Genesis chapters one and two.
📖 Reread *Hurakan and Gucumatz*.
1. Create a key word outline from the first two paragraphs.
2. Retell the story to someone in your own words based on what is in your notes.

L-1c: Ψ Summarizing from Notes
Ψ View IEW Tape 1 - Unit II
Ψ Review IEW pages 9–13: *Summarizing From Notes*

📖 Read *Ra and Sekhmet*
1. Create a key word outline.
2. Retell the story to someone from your notes.
3. ✐ Write the story from your notes.
Note: Always double space.

Extra practice Ψ See IEW page 24

L-1d: Comparison and Contrast – The Flood
📖 Read the following accounts of the Flood:

Civilization	Title	Author or Source
Romans	The Flood	Ovid
Sumerian	The Story of the Deluge	The Epic of Gilgamesh
Greeks	Deucalian's Flood	Greek Mythology
Hebrews	Genesis Chapters 6-8	The Bible

1. Regarded purely as literature, explain which in your opinion makes a better story and why. Consider each of the following components:
   - Action
   - Vividness of presentation
   - Poetic feeling
2. Compare the Sumerian version with that of the Hebrews. Explain the following:
   - Similarities
   - Differences
   - Which predominates, their similarities or their differences?
3. Which version is closer to the Hebrews?
4. Which of the four is most interesting as narrative?
5. Which has the most spiritual quality?

Most scholars feel the Hebrews had borrowed from these various legends.
6. How do you account for any similarities?
7. Why do you think several civilizations have an account of the Deluge?

L-1e: Comparison and Contrast – Eve and Pandora
📖 Read *The Coming of Evil* and Genesis 3
1. Compare and contrast Eve with Pandora
   - Compare: how are they alike or different?
   - Contrast: to show differences by comparing.
2. Increase your descriptive vocabulary and make your stories more interesting and accurate by creating a list of banned words.
   - Use your thesaurus to find five replacement words for each word that you banned.
   - Ψ Refer to IEW page 20.

L-1ee: Be a Movie Critic - *Abraham*
🐚 Watch the movie *Abraham* and answer the following questions with complete sentences and openers.
1. How does Abram show his contempt for the powerless gods of Ur?
2. What do you observe about ancient kings in this video?
3. What does the man in the desert tell Hagar to name her son? What does it mean?
4. How does Abraham solve the problem of the two heirs - the son of his wife and the son of her slave?
5. The sacrifice was made to show Yahweh that you would put nothing of this world above Him. What is Abraham asked to sacrifice?

L-2a: Ψ Sentence Openers

Ψ View IEW Tape 3 – *Sentence Openers*

Ψ Refer to IEW page 19 for notes on sentence openers

1. **Vocabulary** Before reading, write down definitions for these words: quay, omen, succor, rancor, amulet, immolation, lapis lazuli, bitumen, ferrules, and shekel.
2. 📖 Read *The Epic of Gilgamesh*
   - ❑ Read only the summary of Part One
   - ❑ Read Part Two, *"The Forest Journey"*.

L-2aa: Sumerian Writing

The Sumerian form of writing is the oldest form known in human history.

1. Draw examples of pictograph and cuneiform.
2. ✎ Write a few paragraphs summarizing the change from pictograph to cuneiform and its influence on the rest of the world.
   - ❑ Identify the materials used for writing.
   - ❑ When were these translated and by whom?
   - ❑ Use all seven openers in this summary.
   - ❑ Always double space and mark the opener number in the margin.

Encyclopedia keyword: Cuneiform

L-2b: *The Epic of Gilgamesh* Parts 1-2

📖 Reread *The Epic of Gilgamesh* (read the summary of part 1 and all of part 2)

1. List 3 examples of simile, 3 examples of metaphor, and 1 example of symbolism.
2. Write a descriptive paragraph on Gilgamesh and one on Enkidu.
   - ❑ Ψ Refer to IEW Page 10 on *Rewriting and the Use of Computers*
   - ❑ Always double space
   - ❑ Use three sentence openers in each paragraph and note their number in the margin. Use all six sentence openers.
3. What advice was given to Gilgamesh and Enkidu by the city councilors before they left on their journey? Did they follow this advice?
4. Who is the watchman of the woods?
5. What advice did Gilgamesh give to Enkidu before they met Humbaba?
6. Do Gilgamesh and Enkidu epitomize your idea of heroes? Explain.
7. What is the climax of Part Two?
8. Should Gilgamesh have shown mercy to Humbaba? Explain.

L-2c: *The Epic of Gilgamesh* parts 3-4

📖 Read *The Epic of Gilgamesh* parts 3 and 4. Write one to three sentences for each of the following using different openers:

1. What purpose does the repetition within the paragraph on page 30 serve?
2. How did Enkidu's death affect Gilgamesh?
3. Did you see any changes or growth in the main characters as the story progressed?
4. Read the definition of an *epic* in Literature Helps located at the back of the User's Manual. How does this epic express the ideals or nature of Sumeria?

L-2d: *The Epic of Gilgamesh*

Answer the following questions in complete sentences using a different opener for each. Do not use the *subject* opener.

1. Part 2 – Why is Enkidu created?
2. The first dream of Gilgamesh tells his destiny. What does Gilgamesh desire that is not his destiny?
3. Does Gilgamesh need Enkidu to accomplish his destiny?
4. How does Gilgamesh interpret his destiny?
5. Why does Enkidu interpret Gilgamesh's dreams positively?
6. Part 3 – How do the gods behave toward humans?
7. Part 4 – How does Utnapishtim answer Gilgamesh's question, "How shall I find the life for which I am searching?"
8. Everyone in the Garden of the Gods questions Gilgamesh about his appearance--"Why are your cheeks so starved and your face drawn? Why is despair in your heart and your face like the face of one who has made a long journey?" Why do they question him?

L-2dd: Marketing *The Epic of Gilgamesh*

Bonus Design a movie poster for *The Epic of Gilgamesh*. Check the entertainment section of your newspaper or theatres for examples.

L-2e: Book Report

✒ Complete the book report assigned in **L-1aa**. Follow the instructions listed below. Use openers, other than a subject opener for numbers 5 - 13.

1. The title of the book
2. Author
3. ISBN number of book
4. List other books this author has written.
5. Give some of the author's biographical information.
6. What is the theme or moral lesson of this book?
7. Main character's name, physical description and character analysis
8. Name the supporting characters.
9. Time setting as seen in the following excerpt.
10. Geographic setting of this book as seen in the following excerpt.
11. What is the political setting of this book?
12. The major problem in the story.
13. Resolution of this problem.

L-2ee: Be a Movie Critic - *Jacob*

📽 Watch the movie *Jacob*.

Bonus Answer the questions with complete sentences and openers:

1. Why does Rebekah want Jacob to deceive Isaac?
2. How has Esau disqualified himself as leader in Rebekah's eyes?
3. Who is Laban?
4. What is said of Jacob?
5. Consider the law of reaping and sowing. How is this law illustrated in Jacob's life?

L-3a: Hieroglyphics
Of all ancient writings, hieroglyphics are the most commonly known.
1.   Locate the hieroglyphic alphabet on page 34 in the Literature Selections and write your full name.

Use 2 or 3 complete sentences with an opener other than subject opener to answer the following:
2.   What did the Egyptians use hieroglyphics to record and where were they recorded?
3.   What materials were used?

Encyclopedia keyword: Hieroglyphics

L-3aa: Historical Fiction or Biography
📖 Select and begin reading a historical fiction or biography from the Reading Resource List. Other titles are acceptable but you should get approval from your teacher.
※ A book report will be due at the end of Unit 4. ※

L-3b: Egyptian Poetry
The Egyptians wrote proverbs and poetry. Although these poems do not fit the rules of poetry as defined in classical literature, they definitely have the quality of good literature.
📖 Read and enjoy the Egyptian poems provided in the Literature Selections.
1.   Select one of the poems and study the emotion that the author expresses. Read your selection orally with that same emotion to your teacher or classmate.

L-3c: Ψ Dress-ups
Ψ View IEW Tape Two - Dress-ups
Ψ Review IEW pages 17, 18, 20, and 21: Stylistic Techniques
1.   Redo the assignment in **L-1b** with two dress-ups.

L-3d: Tribute
Choose a pharaoh from the Old Kingdom.
1.   Write a tribute that could have been inscribed in his tomb.
   ❑   Your tribute can be poetry or prose.
   ❑   If you select prose, use a different opener in each sentence.
   ❑   Mark the number of the opener used in the margin.
   ❑   Use two different dress-ups in this tribute and underline them.

L-4a: Egyptian Classics –*Story of Sinuhe*
The Egyptians' *Story of Sinuhe* is the oldest known form of novel. It served as a standard educational text for scribes well into the New Kingdom. Many consider it as reasonably factual, though no trace of the real Sinuhe has yet been found. It is the account of an Egyptian courtier, presumably copied from the inscription in his tomb. The story begins with the death of Ammenemes I in 1961 BC (12th Dynasty) as reported by his son and successor Sesostris I.

📖 Read the *Story of Sinuhe* and answer the following:
1. List examples of propaganda from this story. See page 4 of Literature Helps.
2. Review Story Elements from Literature Helps. List these elements for the *Story of Sinuhe*. Are the characters flat or round?
3. Does this story follow all aspects of a good story? Discuss and justify in a paragraph. Use openers and two dress-ups.
4. The narrator flees from the court though no explanation is given. He meets a traveler who greets him respectfully, "for he was frightened". What can you speculate from these incidents in the story?

L-4aa: **Bonus** Be a Movie Critic - *Joseph*
🎥 Watch the movie, *Joseph* (PG-13) produced by Turner Home Entertainment
1. List the story elements.

Note to Teachers: This PG-13 movie covers mature topics, e.g. the seduction of Joseph by Potiphar's wife, the rape of Dinah, and the prostitution of Tamar. Refer to Genesis chapters 34, 37 - 46 to get an understanding of the material that is covered in the film.

L-4b: Ψ Decorations
Ψ View IEW Tape 4 - Decorations
Ψ Read IEW page 22: *Examples of Decorations*

L-4c: Egyptian Classics – *The Tale of the Doomed Prince*
📖 Read *The Tale of the Doomed Prince*. The theme, like *The Story of Sinuhe*, is of an Egyptian that is abroad. We can see a degree of flexibility in the Egyptian idea of predestination through the three possible fates that may prevail on him. Unfortunately, the conclusion is lost.
1. Considering what you have learned about the Egyptians of this age, their religion and beliefs, complete the story giving it the ending you desire.
   - As always, double space.
   - Use all openers and mark their number in the margin.
   - Include three dress-ups.
   - Use two decorations putting a dotted line under them and writing DEC in the margin.

L-4d: Interview with a Pharaoh
Pretend you are a reporter for the magazine of your choice. You have been given the following assignment: *Interview one of the major pharaohs (Hatsheput, Iknaton, Tutankhamen, Ramses II)*. Your interview must include at least six questions. Careful attention must be given so that your questions reflect topics in which your magazine's readers are interested. Write your questions and answers in a script format or, better yet, videotape the "actual interview".

L-4e: Book Report
    ✐ Complete the Book Report assigned in **L-3aa**. Use the critique outline to write your report.
    Note: If not using IEW with *Discovering the Ancient World* use questions from **L-2e**.
    Ψ View IEW Tape 6: Unit IX Critiques
    Ψ Review IEW pages 67 – 71

L-5a: Phoenician Script
    1. Research and find out what the purpose for written language was for the Phoenicians.
    2. What became of the Phoenician script?

    Write two or three sentences using openers, two dress-ups, and one decoration with appropriate markings.

L-5b: Historical Fiction or Biography
    📖 Select and begin reading a historical fiction or biography from the Reading Resource List. Other titles are acceptable but you should get approval from your teacher.
    ※ A book report will be due at the end of Unit 6. ※

L-5c: Monologue: My Life as a Phoenician
    After studying the Phoenicians, choose a character to develop a persona around: ship builder, sailor, craftsman, child, wife, ruler, etc. From your study speculate what life would be like for this person. Give him/her a name and write a biographical sketch. Using this information, prepare a monologue in which you portray this fictitious personality. You will need an audience and be in costume for your presentation. You will be graded on the following points, five points each:
    1. Background of the character
    2. Written in first person
    3. Setting of character given in an interesting manner
    4. Life of the character given in chronological order
    5. Other important historical figures mentioned in dialogue
    6. Monologue given in the verbiage appropriate to the character
    7. Give the characters prevailing motivation
    8. Character's historic and political significance in the main point
    9. Character's personal life
    10. Character's major contribution to history and society given at the ending.
    11. Humor or story added in appropriate historical setting
    12. Monologue engaging
    13. Monologue 3 to 5 minutes long
    14. Gestures
    15. Eye contact in presentation
    16. Note card with 50 or less words
    17. No notes (10 points). This option provides an extra five points.
    18. Show minimum of 20 hours work on dress, monologue, and written paper
    19. Paper is typed, double spaced and in Times New Roman
    20. Paper should have a heading and title
    Note: If you practice in front of a mirror it will help you with your presentation.
    Ψ Refer to IEW page 8 for tips on public speaking.

L-5d: Phoenicians at the time of Ezekiel
Bonus Ezekiel wrote of the Phoenicians.
1. What did Ezekiel say about the Phoenicians? List references.

L-6a: Crete – Home of a Lost Civilization
The ancient civilization of Crete went unnoticed for nearly 2000 years but was the setting of more than a few Greek Myths. The Greek god Zeus was born in a cave high atop a mountain on Crete. His son, Minos, ruled Crete. A creature half man, half bull was held deep inside King Minos' palace at Knossos. The Minotaur was held in the heart of a dark labyrinth built by the famous mythical architect, Daedalus. It was these legends and more, rich in tradition but elusive historically, which drew famed archaeologist Sir Arthur Evans to Crete. He was a wealthy man who bought a mound that was thought to be the legendary site of Knossos. In March of 1900, myth leaped to life as the dig began which eventually unearthed a magnificent palace. Gradually, an extraordinary civilization was revealed which Evans named "Minoan" after the legendary King Minos. Evidence of the Minoan's written language referred to as Linear A and Linear B have been found. Linear A has yet to be deciphered. Michael Vintris deciphered Linear B in 1954.
1. What is known of these scripts?
2. What does it appear that they were used for?

L-6b: Greek Mythology – *Theseus and the Minotaur*
When reading Greek mythology it is important to be aware of some of the things you might come across. Greek myths are full of monsters, magic, and unnatural relationships. You will read two of the legends that drew Sir Arthur Evans to excavate on Crete.
📖 Read the story of *Theseus and the Minotaur.*
1. Note any references that gave Evans clues in his search for this ancient civilization.

L-6c: Greek Mythology – *Daedalus and Icarus*
The study of Greek myths shows a distinct family tree of the gods and a continuing epic of events in the heroes' lives.
📖 Read the story of *Daedalus and Icarus.*
1. How does the story of *Daedalus and Icarus* seem to relate to the story of *Theseus*?
   ❑ Justify your answer in a paragraph.
   ❑ Use all six openers, two dress-ups, and one decoration. Always double space.

L-6d: News Analysis – *The Minoans*
Pretend that you are a reporter for a major news magazine. Your assignment is to write a story speculating on what happened to their civilization. After studying the Minoan civilization write your news story keeping in mind the following:
- ❑ News reporters always answer the following questions in the *lead* (first) paragraph: who, what, when, and where.
- ❑ News reporters write using the third person. This refers to using pronouns such as he, she, it, or they when referring to the person or thing spoken about.
- ❑ Use a catchy headline that captures the attention of the reader. Don't forget your byline.
- ❑ Use openers, two dress-ups, and one decoration with all the appropriate markings.

For reference, look at examples in your newspaper. Use a highlighter and mark the lead paragraph. Does the reporter answer all the questions?

L-6e: Book Report
✒ Complete the Book Report assigned in **L-5b**. Use the critique outline to write your book report.
Note: If not using IEW with *Discovering the Ancient World* use questions from **L-2e**.
Ψ Review IEW Tape 6: Unit IX Critiques
Ψ Review IEW pages 67 – 71

L-7a: Chinese Folklore
📖 Read the *Golden Reed Pipe* and the *Vulnerable Spot*. These stories date as far back as the Shang Dynasty (pronounced Shong) and were not necessarily written for children.
1. What type of stories would these be classified as today?
2. Do they remind you of any childhood stories you have ever read? Which ones?
3. List the elements of these stories.
4. How do they compare in style and content with the stories from Egypt?

L-7aa: Historical Fiction or Biography
📖 Select and begin reading a historical fiction or biography from the Reading Resource List. Other titles are acceptable but you should get approval from your teacher.
※ A book report will be due at the end of Unit 8. ※

L-7b: Ψ Summarizing Narrative Stories
Ψ View IEW Tape 2: Summarizing Narrative Stories
Ψ Review IEW pages 27-36
📖 Reread *The Golden Reed Pipe*
1. Ψ Write a summary using IEW pages 27-36 as a reference.
Extra practice Ψ IEW pages 24, 28, 32-33

L-7c: Creative Writing
Ψ View IEW Tape 5 – Unit VII
Ψ Read IEW pages 53 – 55 *Creative Writing* and study the examples.

During the latter period of the Chou Dynasty the Golden Age of Chinese philosophy emerged. There were three main schools of philosophic thought that appeared: Confucianism, Taoism, and Legalism. Originally these philosophic ideas were to improve the leadership and government.

1. Briefly summarize each philosophy.
2. Do you see any of these philosophic ideas in our society or government today?
3. ✐ Write a paper on Concepts of Chinese Philosophic Ideas Seen in Government Today.
   ❑ You will need to mention two concepts of thought from Confucius, Taoism or Legalism. Tell how this is seen as a concept used by governmental leaders today. One such concept is if the leader sets a good example his people will imitate him. You are presenting facts only in the body paragraphs, but the conclusion may be your own opinion concerning this concept, whether it is a Biblical concept, a moral decision, or an ideal we should follow or not.
   ❑ Remember; do not use the word "I" when writing the conclusion. Re: Ψ Page 67, #4
   ❑ This paper should have an introduction, two body paragraphs and one concluding paragraph.
   ❑ Each paragraph should begin with a different sentence opener. Write the number of the opener in the margin.
   ❑ Each paragraph should use a different dress-up. Underline all dress-ups.
   ❑ Each paragraph should use a decoration and label.

Your paper will be evaluated on:
1. Style – Use sentence openers, dress-up, and decorations.
2. Content – Can you quote the philosopher?
3. Use of excerpts from their writings – Identify excerpts with quotation marks.
4. Your opinion – Can you substantiate what you think?

L-7d: Confucius Says
📖 **Bonus** Read from the writings of Confucius or Lao Tsi.
1. How do these writings compare with Solomon's or even the words of Christ?
2. Present your comparison in a paragraph using all openers, two dress-ups, and one decoration with all the appropriate markings.

L-7e: As Smooth as Silk
📖 Read the legend of how silk was discovered - *The Legend of Hsi-Ling-Shi*, bride of Emperor Huang Ti. There are many versions of how she discovered the thread that made the cocoons.
1. How do you think this discovery resulted?
2. Write your own version of this legend.
   ❑ Use all openers, two dress-ups, and one decoration with all the appropriate markings.

Encyclopedia keyword: silk

L-8a: Ψ Triple Extensions
Ψ View IEW Tape 4
Ψ Read IEW page 23: Triple Extensions.
1.   Write example sentences for all six triple extensions.

L-8b: The Rig Veda on the All-Maker
The *Rig Veda* is to the Aryans or now Hinduism as the *Torah* is to the Jews and the *Bible* to Christians. The *Rig Veda* is a text of 1028 hymns, each averaging about ten verses. It is one of a larger collection divided into four groups: *Sama Veda*, *Yajura Veda*, and *Atharva Veda*. These originated with the Aryans sometime between 1300 and 1000 BC. The Vedic texts are primarily ritual handbooks, a type of liturgy, consisting mainly of hymns, sacrificial formulas, incantations, and spells. They are revered as "apaurusheya" – *not of human origin.*

The *Rig Veda* explores many different theories of creation, from the results of a cosmic battle to the unmotivated act by the gods of separating the heavens and the earth. The Hymn 10.81 refers to the All-Maker and speculates on the mysterious period surrounding creation.
1.   How is the All-Maker described? Write your answer as a descriptive paragraph.
   ❑   Use all openers, two dress-ups, and a triple extension with the appropriate markings.

L-8c: The Rig Veda on Death
Just as the *Rig Veda* speculates several variations of creation, so it has much speculation about death. They practiced both cremation and burial. They had several ideas concerning the fate of the dead. They wrote of heaven, a new body, revival, reincarnation, and dispersion among the elements. Many people may be addressed in a single hymn: dead ancestors, the gods, mother earth, death, the dead man, as well as the mourners. Death was regarded with great sadness but without fear.

Hymn 10.135 centers on a young boy whose father has died. The boy mentally follows his father's journey to the realm of Yama who is the king of the dead. Use openers, two dress-ups, and a triple extension with appropriate markings answering all the following questions.
1.   What do you think the chariot in verses three, four, and five symbolizes?
2.   What message do you think the author is trying to give?

A great emphasis is placed on sacrifice in the *Rig Veda*, not just physical or ritual sacrifice, but the sacrifice of speech, faith, praise, and generosity, just to name a few. Hymn 10.117 exhorts the worshipper to be generous to the gods, the poet (priest), as well as to mankind in general. This hymn also gives advice to the one who would be generous and cautions the one who would not.
3.   Explain the theme of this hymn.
4.   Write a paragraph comparing this hymn to teachings from the Bible.
   ❑   Include an excerpt to justify your position on your comparison.

L-8d: The Bible and Wisdom
Bonus Contrast the wise and the foolish as presented in the *Bible*.
1. List scriptures from the *Bible*, their references, and a summary of each verse concerning the wise and the foolish.
2. Sum up what you have learned in an expository essay, presenting information that explains your analysis to the reader.
    ❑ Use all openers, two dress-ups, one decoration, and one triple extension with the appropriate markings.

L-8e: Book Report
🖊 Complete the book report assigned in **L-7aa**. Use the critique outline to write your report.
Note: If not using IEW with *Discovering the Ancient World* use questions from **L-2e**.
Ψ View IEW Tape 6: Unit IX Critiques
Ψ Read IEW pages 67 – 71

L-9a: Ancient Mesopotamian Literature
Gaining and understanding of the literature of the ancient Mesopotamian region can be a frustrating and confusing experience. Scholars spend years studying the background, society, culture, and religion of the ancients before attempting to analyze the translations of this period. One thing you need to keep in mind is that this region was inhabited by many civilizations; the myths, legends, and even religion of each civilization were absorbed into the next. This is particularly clear with the *Epic of Gilgamesh,* which originated with the Sumerians. Versions of this epic have been found in the literature of the Hurrians, Hittites, Old Babylonians, Assyrians, and Akkadians. So, in reading the Old Babylonian poetry, scholars are not at all certain where it originated.
The poem you are about to read, *A Prayer to the Gods of Night*, is one such poem. As you have learned, the religion of Old Babylon was polytheistic.
1. Write an essay on what you can learn of their gods and beliefs through this poem.
    ❑ Justify your answer with excerpts from the poem.
    ❑ Use all openers, two dress-ups, one decoration, and one triple extension with the appropriate markings.

L-9aa: Historical Fiction or Biography
📖 Select and begin reading a historical fiction or biography from the Reading Resource List. Other titles are acceptable but you should get approval from your teacher.
※ A book report will be due at the end of Unit 10. ※

L-9b: The Code of Hammurabi

The code of Hammurabi, inscribed on a seven and one half foot stele, contains judgments and legal decisions of King Hammurabi. One of the most important duties of an ancient Semitic ruler was to settle disputes. The stele of Hammurabi is a collection of decisions in cases that had been brought before him. The "laws" or "codes" consist of about 280 sections sandwiched between a prologue and epilogue. The prologue claims that the gods called Hammurabi "to make justice visible in the land, to destroy the wicked person and evildoer, that the strong might not injure the weak". The epilogue explains the purpose for the inscriptions, "to set right the orphan and widow." It counsels the oppressed with, "Let any oppressed man who has a cause come into the presence of my statue as king of justice, and have the inscription on my stele read out, and hear my precious words, that my stele may make the case clear to him; may he understand his cause, and may his heart be set at ease!" Hammurabi's reign seems relatively humane in comparison to other civilizations of this time. Some scholars liken him to Solomon.

📖 Read the excerpts of his code from Literature Selections

1.   Choose one code and compose a short story narrative illustrating it.
   ❑   Use all openers, three dress-ups, one decoration and one triple extension with all the appropriate markings.

L-9c: The Code of Hammurabi and the Bible

Bonus Compare the teaching of the Bible with the Code of Hammurabi.

1.   Find a scripture from the Bible for each code listed. The scripture may agree with or oppose the code selected. Write both the code and the relating scripture and reference.

L-10a: Homer's *Iliad*

The city of Troy and the civilization at Mycenae remained legend until 1870 when a German archaeologist decided to dig the site recorded in Homer's *Iliad* of the Trojan War. He found evidence of Troy and a war that resulted in its destruction, hence the Mycenaean Empire. The Trojan War lasted ten years; the *Iliad* deals with only a few months of the war.

1.   Look up and define: hecatomb, felicity, whelp, tactician, emissary, faggot, emulous, revoke, ambrosia, carnage, and rogue.
2.   Read and write a summary of the following books from the *Iliad*: books one, two, sixteen, and twenty-four.
   ❑   Use all openers, two dress-ups, one decoration, and one triple extension. Double space and use all appropriate markings.

The following is a brief chapter outline to help give you the full picture of what takes place in the *Iliad*. A list of the characters is included in the Literature Selections. Notice that the ancient Greeks are referred to as Achaians.

Chapter	Description
3	A duel and then a truce take place between Meneloas and Paris
4	Pandaros breaks the truce
5	Diomedes (a hero) fights the gods
6	Hektor returns from the battlefield to visit his family
7	Ajax duels with Hektor
8	Zeus influences the battle
9	Achaians send emissaries to Achilles
10	A foray takes place at night in camp
11 to 16	Trojan successes
17	Fight for Patroklos' body
18	Achilles gets new armor made by the gods
19	Achilles rejoins the Achaians
20	The gods go to war
21	Achilles battles the god of the river
22	The death of Hektor
23	Funeral games for Patroklos

L-10b: Homer's *Iliad – The Characters*
1. List three positive and three negative qualities you observe in Achilles, Agamemnon, Hektor, Patroklos, Priam, Zeus, and Apollo. You may have to infer these qualities from their behavior.
2. Using this list, write an essay on the character qualities of the various personalities. In paragraph one note the virtues of the characters. In the second paragraph note their negative aspects.
   - Use all openers, three dress-ups, one decoration, and one triple extension. Double space and use all appropriate markings.

L-10c: Homer's *Iliad* – Q & A
Answer the following questions with complete sentences.
1. In Book One an argument arises between two military leaders.
   - What are their names?
   - What is the nature of their dispute?
   - Are they fighting against the same enemy?
   - Who wins the argument and what does he do to show his superiority?
   - How does Achilles take revenge?
3. When the army reunites in Book 16 there is a scene where the gods decide the fate of Sarpedon.
   - Who is Sarpedon's father?
   - What arguments convince the father to allow his son to be killed?
   - Of all the gods, who is most feared, honored, and called upon to make the final decisions?

4. In Book 24 we see opposing facets of Achilles' character.
   - ❑ Give an example in which his heart is hard and bitter and then in which he becomes kind and understanding.

L-10d: Homer's *Iliad* – Historical Significance
The introduction of the two armies in Book Two has been of great value to historians and archaeologists.
1. Explain how in a paragraph with openers, decoration, dress-up, and triple extension.

L-10e: Book Report
✐ Complete the book report assigned in **L-9aa**. Use the critique outline to write your report.
Note: If not using IEW with *Discovering the Ancient World* use questions from **L-2e**.
Ψ View IEW Tape 6: Unit IX Critiques
Ψ Review IEW pages 67 – 71

L-10ee: Genealogy – …And Grandma Beget Mom, and Mom Beget Me.
Bonus Lineage was very important to the ancients.
1. Write your own genealogy back four generations.

L-11a: Lost Civilizations
1. Explain why there is no written script preserved from the Olmec and Nubian civilizations.

L-11aa: Historical Fiction or Biography
📖 Select and begin reading a historical fiction or biography from the Reading Resource List. Other titles are acceptable but you should get approval from your teacher.
※ A book report will be due at the end of Unit 12. ※

L-11b: Essay Writing
Ψ View IEW Tape 4
Ψ Read IEW pages 53-55: Essay Writing
1. Using the "Champlain" model as an example, write an essay comparing the Olmec with the Egyptians.
   - ❑ Consider their government, writing, architecture, agriculture, and physical features.
   - ❑ Use at least three items of comparison from the civilization questionnaire.
   - ❑ Use three openers, two dress-ups, one decoration or one triple extension with appropriate markings per paragraph.

L-11c: Nubian Wisdom
Having studied the lifestyles of the Nubians:
1. What do you think is the meaning of this Nubian proverb, "One man cannot build a home, but ten men can easily build twenty homes."
   - ❑ Write your opinion in a paragraph using openers, two dress-ups, one decoration, and one triple extension. Double space and use all appropriate markings.

L-11d: Nubian Commerce

The Nubians were merchant middlemen for trade between Egypt and the interior of Africa. Take a modern invention and think of a way to market it to these ancient people. Remember, there is no electricity, batteries, or gasoline.

1. Write a script for a "commercial" that could be acted out in the market square to sell this amazing "must own" product.

    ❑ Your script should have dialogue and stage directions. Dialogue is what the actors will say. Stage directions are usually written in italics and tell how the actors will look, move, and speak. They also describe the setting, sound effects, and lighting.

L-12a: Short Story – Life in Israel

Your assignment is to write a short story or part of a novel. This will be due at the end of the Units on the Hebrews **L-13e**.

To turn in:

   ❑ A story outline
   ❑ Character sketches
   ❑ A rough draft
   ❑ A finished and typed short story
   ❑ Illustrations are beneficial but must be in addition to text.

Format specifications:

   ❑ One-inch margins
   ❑ Font: 12 point Times New Roman
   ❑ At least five pages of text in length

Story Components

   ❑ Your story takes place in Jerusalem during the reign of King Solomon or King David.
   ❑ Your story must take place within two days time.
   ❑ Your story should include no more than three locations within the city.
   ❑ You may have up to three characters, at least one being a woman. These can be real people or fictional.
   ❑ You must include an animal.
   ❑ Your story must be historically accurate.

You will need to research the time frame to make your story historically accurate. The Bible is a good resource. You will need to consider what they ate, what they wore, how they slept, what animals were found or kept in ancient Jerusalem, what things concerned them, etc. Whatever details needed to be included in your story. If you select a real individual to portray, research them carefully.

Even though this has a historical setting, write about something you know about.

Writing your Story Outline
I. Beginning
    A. Set the scene
    B. Introduce main character
    C. Present the problem
II. Middle
    A. Tell what the main character does
    B. What happens?
III. Ending
    A. How does the main character solve the problem?
    B. How does the story end?

ΨYou may also use the outline in IEW page 27.

Writing a Character Sketch
  ❑ Describe their appearance: tall/short, hair and eye color, clothing, age
  ❑ Describe their family: parents, siblings, extended family
  ❑ Describe their personality: happy/sad, shy/outgoing, passions/fears
  ❑ Describe their surroundings: house, neighborhood, city
  ❑ Describe their interests: sports, hobbies, pets, school
  ❑ Describe anything you will need to know about this person to tell the story.
  ❑ Your task will be easier if you write a sketch for each of your characters.

Select a theme for your story; what do you want to say to your reader?
Select the point of view. Who is telling the story? The easiest way to write your story is to have one main character with the story told through their eyes.
Use a balance of dialogue and narration. Dialogue makes the story more dynamic. Select words that describe feelings, emotions, and appearance. Use your thesaurus for more variety.

L-12aa: Be a Movie Critic - *Moses*
    ❧ Watch the movie, *Moses*
1. Write a Movie Review
    ❑ Ψ Refer to the example in IEW page 70 as a guide.

L-12b: Old Testament Literature
The Old Testament is rich in literature; in it are found biography, history, short story, dramatic narrative, poetry, and proverbs. Of all ancient literature that deals with historical accounts, most scholars regard the Old Testament as fairly accurate and an objective representation of history.

The Pentateuch, Joshua, and Judges were recorded during this period of Israel's history. Moses is generally regarded as the original author of the Pentateuch. However, the original manuscripts have never been found. It is held that the stories of Genesis were passed down by oral tradition until Moses undertook writing them down. Also the death of Moses recorded in Exodus is considered to have been appended later. The authorship of Joshua is unknown, although Jewish tradition attributes it to Joshua; and later appended to by Eleazar and Phinehas. The book of Joshua gives account of Canaan's successful conquest by the Israelites. Jewish tradition assigns the authorship of Judges to Samuel but this has not been proven. Whoever it was, the author seems to have obtained his material from diverse sources, both oral and written.

📖 Review the first seven books of the Old Testament and determine what type of literature it is.

Note: Some may have examples of several forms of literature included in them.

L-12c: Old Testament Songs of Victory
📖 Read Exodus 14, 15: 1 - 18, and Judges 4-5. Both of these passages include victory songs.
1. Contrast the *Song of Moses* with the *Song of Deborah*.
   ❑ Ψ Use the form from IEW pages 53 – 55 as your guide.

L-12d: Short Story – Checkpoint
Your story outline, research, and character sketches should be complete for assignment **L-12a**.

L-12e: Book Report
✒ Complete the Book Report assigned in **L-10e**. Use the critique outline to write your report.
Note: If not using IEW with *Discovering the Ancient World* use questions from **L-2e**.
Ψ View IEW Tape 6: Unit IX Critiques
Ψ Review IEW pages 67 – 71

L-13a: Old Testament Psalms and Proverbs
The Psalms and Proverbs are probably the most notable of Hebrew Literature. Authorship is varied with David, Solomon, Moses, Asaph, and others who are anonymous contributing. The themes vary as well, from personal psalms of distress and prayer, historical and Messianic psalms, to psalms of worship and praise to God.

The form of Psalms is free verse, relying on repetition and balance rather than rhyme and meter for its poetic feeling. Although originally written in prose, later translations printed them as poetry to aid in hearing the rhythm. Common rhythmic devices used in the Psalms are:
   ❑ *Repetition of idea* – as in Psalm 20, "May he <u>send you help from the sanctuary, and give you support from Zion!</u>" – the idea is the same with different phrasing.
   ❑ *Refrain* – as in Psalm 118 that repeats, "His steadfast love endures forever". Refrains usually run in pairs but sometimes in longer series.
   ❑ *Contrast* – as between despair and hope in Psalm 43.
   ❑ *Questions and Answers* – as in Psalm 43, "Why are you cast down, O my soul?" "Hope in God; for I shall again praise him".

A single device may be used or a combination, (as seen in Psalm 43), or two or more throughout the Psalm. Simile, metaphor, and symbolism are used throughout the Psalms.

📖 Read Psalm 1, 5, 24, 70, and 136 then answer the following:
1. What metrical devices, as explained above, were used in each?
2. What are the mood and theme of each?
3. Select three examples of figures of speech and relate what they are describing.
4. Choose one of the Psalms to memorize and say to your teacher or class.

L-13aa: Historical Fiction of Biography
📖 Select and begin reading a historical fiction or biography from the Reading Resource List. Other titles are acceptable but you should get approval from your teacher.
※ A book report will be due at the end of Unit 14. ※

L-13b: Old Testament and Ancient Civilizations
Bonus Many of the ancient civilizations you have studied (and some you will study) are mentioned in the Old Testament.
1. See how many you can find.

L-13bb: Short Story – Rough Draft
✐ Write your rough draft and set it aside for a few days.

L-13c: Paraphrasing and Summarizing
Paraphrasing is putting the writer's meaning into your own words. Summarizing requires you to restate it briefly in different words.
1. Paraphrase Psalm 1 and write a descriptive summary.
   ❑ Use openers, dress-ups, two decorations, and one triple extension. Double space and add appropriate markings.

L-13d: Short Story - Editing and Giving Your Story a Title
Ψ Review IEW page 14: General Style Guidelines
📖 Reread your story.
   ❑ Test the flow and content by asking the question "why". The more times you can ask and answer this question the more complete the story.
   ❑ Are your facts correct?
📖 Read your story aloud.
   ❑ Does the dialogue sound like people talking?
   ❑ Do you repeat words or phrases?
   ❑ Are your descriptions and characters consistent?
Update your Banned Words list.
   ❑ Use your thesaurus for descriptive words and variety.
Check for spelling and punctuation errors.
Select your title.
   ❑ It can be something that describes or calls attention to a character, the plot, or theme.
   ❑ It can be a clincher – refer to Ψ page 29, #6

L-13dd: Short Story - Type your story.
Type your story keeping in mind the Format Specifications given in **L-12a**.

L-13e: Short Story – Completion
Design a cover for your short story in an "ancient" style. It should include:
- ❑ A story summary
- ❑ Author's bio
- ❑ Critical comments – Use actual quotes from people who have read your story.

L-14a: Homer's Odyssey – Did Homer Exist?
Opinions concerning the true existence of Homer have changed over the past thirty or forty years with various explanations depending on your source. Most critics agree that the *Odyssey* was composed in the eighth or ninth century BC and were passed down by oral tradition until the sixth century BC when, in Athens, they were written down for the first time.
1.   What are the current ideas concerning Homer's life and career?

L-14b: Homer's Odyssey - Wandering
Odyssey is a Greek word that means "the tale of Odysseus". The *Odyssey* is concerned with the adventures and exploits of Odysseus, King of Ithaca and a hero of the Trojan War. When the *Odyssey* begins, ten years have passed since the fall of Troy and Odysseus still has not returned home. In his absence, his palace has been taken by various nobles hoping to woo Penelope, his wife. She, however, has remained faithful and will not remarry. Telemachus, the son of Odysseus, visits the Achaian lords who have returned trying to find news of his father. During this ten year period Odysseus had many adventures and encountered many dangers. The selection you will read focuses on a portion of Odysseus' wanderings. In Book IX Odysseus has just arrived on the island of Scheria, the home of the Phaeacians, where at the palace of the king he reveals his identity and begins to recount his adventures. Upon finishing his story the generous Phaeacians give Odysseus many gifts and transport him safely back to Ithaca. Once home Odysseus takes vengeance on the suitors and re-establishes himself as king.
- ❑ **Vocabulary**: vermilion, barque, hawser, whey, wain, enmity, and guile.

📖 Read Book IX of the Odyssey and answer the following questions.
1.   Many of Odysseus' qualities are revealed in this narrative. Which lines show his character?
2.   What passages tell something about the customs, activities, culture, etc., of the ancient Greeks?
3.   As in the *Iliad*, we see the gods playing a vital role in the experiences of the humans. What attitude do the characters show toward the gods? What attitude do the gods show toward the humans?
4.   The *Odyssey* is an example of episodic plot. List other examples of literature we have studied that have an episodic plot.

L-14c: Homer's Odyssey - Questions
Answer the questions concerning the Odyssey using complete sentences.
1. How does Odysseus introduce himself to King Alcinous?
2. At Ismarus, Odysseus' boast that they escaped death and destiny. What happened there?
3. What happened to those who ate the honey-sweet fruit of the lotus?
4. What is the occupation of the Cyclopes?
5. Why do the Cyclopes pay no heed to the gods?
6. Why doesn't Odysseus kill the sleeping Cyclops and avenge his companions?
7. What is the stranger's gift that Polyphemus promises to Odysseus?
8. How does Odysseus' selected name 'Noman' protect him and his companions?
9. How did the saying; "Forewarned is forearmed" fail Polyphemus in the case of Odysseus?
10. What does Polyphemus pray concerning Odysseus?

L-14d: Homer – Author or Legend
Having read portions of the *Iliad* and *Odyssey* do you agree or disagree with critics concerning Homer's authorship?
1. Write your opinion in a paragraph. Use openers, two dress-ups, one decoration, and one triple extension. Double space and use appropriate markings.

L-14e: Book Report
✎ Complete the Book Report assigned in **L-13aa** using the questions from **L-2e**.

L-15a: Jonah's Message To Nineveh
Assyria began to rise as a world power about 900 BC, near the close of King Solomon's reign and the division of Israel. The Assyrians gradually absorbed and destroyed the Northern Kingdom of Israel.

Jonah came from Gath-hepher in the northern part of Israel. His ministry, during the reign of Jeroboam II (790 - 749), coincided with the ministries of the prophets Amos and Hosea. Jonah was sent to Nineveh by God to persuade them to repent, to prolong the life of "the enemy nation" that was probably the most brutal yet studied.

Naturally, the unbelieving mind finds it hard to accept the story of Jonah. It is called a parable, an allegory, or fiction, but Jesus regarded it as historic fact (Matthew 12:39-41, Luke 11:29-32). He compared Jonah's experience to his own death, burial, and resurrection. No archaeological evidence has yet been found of Nineveh's repentance although archaeologists do note reformation during Adad-Nirari's reign. During his reign and that of the three following kings, there was a reprieve of Assyrian conquests. During this period of reprieve Israel recovered lost territory.

What was so powerful in Jonah's words to cause a nation to repent? The Old Testament does not go into detail about what Jonah said.
✎ Write a sermon for Jonah.
❑ In your sermon explain what you think he could have said.
❑ Use devices in your sermon such as allegory or comparisons, and metaphors that would have made Jonah so believable.
❑ Consider his experience and the religious beliefs of the Assyrians and include in your sermon.

L-15b: Jonah's Arrival In Nineveh

A feature story is a human-interest story based on news. It is of public interest rather than news information. Locate a feature story in your newspaper and read it. Observe how the reporter weaves the facts with feelings.

1. Write a feature story about the arrival of Jonah or his sermon and the crowds' reaction.
    - ❑ Use a catchy headline that captures the attention of the reader.
    - ❑ Don't forget your byline.
    - ❑ Use openers, two dress-ups one decoration, and one triple extension with all appropriate markings.

L-15c: Ψ Writing from Pictures

Ψ View IEW Tape 4

Ψ Read IEW pages 47-52: Writing from Pictures

1. Write a three-paragraph story from the pictures included in the Literature Selections, page 158.

L-16a: Etruscan Script

Only fragments of the Etruscan script have been saved.

1. Write a paragraph describing what these consist of.
    - ❑ Use openers, two dress-ups, one decoration and one triple extension with all appropriate markings.

L-16b: Ψ Summarizing References

Ψ View IEW Tape 3

Ψ Read IEW pages 37-46: Summarizing References

✎ Write a paper on the Etruscans.
    - ❑ Use three topics, three paragraphs composition.
    - ❑ Consider such topics as religion, art, life style, and their influence on the Romans for your topics.

L-16c: Ψ Personal Letters

Ψ View IEW Tape 3

Ψ Read IEW page 57: Example of a Personal Letter

1. Write a personal letter in first person, as if you were an Etruscan, to a friend describing its decline as a powerful civilization.
    - ❑ Take into account the superstitions, beliefs, and lifestyle of the Etruscans.
    - ❑ Include the factors that contributed to their collapse.

L-17a: Short Story – Life in Babylonia

Refer to the short story assignment in Unit **L-12a**. Use the same timetable for this assignment. It will be due the end of Unit **L-18**.

✐ Write a short story

- ❑ Your story takes place in ancient Babylon during the reign of King Nebuchadnezzar II.
- ❑ Your story takes place within two days time.
- ❑ Your story should include no more than three locations within the city.
- ❑ You must have three characters at least one being a woman.
- ❑ You must include an animal or a natural disaster.

Note: The Bible is good resource for a description of life in Babylonia.

L-17b: World View

1. **Define** "world view".
   - ❑ What is a *secular* world view?
   - ❑ What is a *Christian* world view?
   - ❑ Compare a Christian world view to a secular world view.
2. Write a persuasive essay on your own world view.
   - ❑ Ψ Refer to IEW page 63 for a model.
   - ❑ Give examples of your own world view through your choices of art, architecture, music, movies, and literature.
   - ❑ Explain how your philosophy of what you believe -- "your world view" aligns with your choices in the activities of your life?
   - ❑ Use openers, two dress-ups, and a decoration or triple extension with the appropriate markings.

L-17c: World View – King Nebuchadnezzar

📖 Read chapters one through four from the Old Testament book of Daniel.

1. What was Nebuchadnezzar's world view before his dreams?
   - ❑ Sum up in a paragraph his world view before the dream.
   - ❑ Write a paragraph illustrating how his experiences changed his world view?
   - ❑ Use openers, two dress-ups, and a decoration or triple extension with the appropriate markings.

L-17d: Book Summary – Daniel

Summarizing is restating briefly in different words the main ideas of a passage.

1. Summarize the book of Daniel in the Bible and provide the theme.
   - ❑ Use openers, two dress-ups, and a decoration or triple extension with the appropriate markings.

L-17dd: Comparing Bible Characters

Daniel wasn't the only Hebrew to be elevated to the highest levels of a non-Hebrew government.

1. What other Hebrew have you studied that was raised to the right hand of the ruler?
   - ❑ Explain the similarities

L-17e: World View - Debate
Bonus Select teams and debate different aspects of world view.
- Ψ Refer to IEW page 63 for a debate format.
- Select the themes you will debate and be knowledgeable of both sides of the argument.
- Present your argument with evidence and logic.

L-17ee: Short Story – Checkpoint
Your story outline, research, and character sketches should be complete for assignment **L-17a**.

L-18a: Scythian Origin Myth
📖 Read the Scythian Origin Myth
📖 Read Genesis 12:1-7
1. Compare the myth with the idea of selection in Genesis.
- Write a paragraph using openers, two dress-ups, and one decoration with appropriate markings.

L-18b: Comparison of Symbols
Select a symbol from one of the previously studied civilizations and write a paragraph comparing it to the Scythian symbol of the golden bowl.
- Use all openers, two dress-ups, one decoration and one triple extension with all appropriate markings for this comparison.

L-18bb: Short Story – Rough Draft
✎ Write your rough draft and set it aside for a few days.

L-18c: Amazon Women
The Amazon legend comes from the Greeks. Aeschylus calls them "The warring Amazons, men-haters". They were a nation of women, all warriors. They were supposed to live around the Caucasus and their chief city was Ahemiscyra. Curiously enough, they inspired artists to make statues and pictures of them far more than poets to write of them. Familiar though they are to us, there are few stories about them. They invaded Lycia and were repulsed by Belerophon. They invaded Phrygia when Priam was young, and Attica when Theseus was king. He had carried off their queen and they tried to rescue her, but Theseus defeated them. According to a story not in the *Iliad*, they fought the Greeks in the Trojan War under their queen, Penthesilea. They were allies of the Trojans. Pausanias says that she was killed by Achilles, who mourned for her as she lay dead, so young and so beautiful.

Herodotus of Halicarnassus has been called "the father of history." He lived during the height of Classical Greece. He spent most of his life traveling and studying the people he met. He then recorded what he had learned. He says that he writes "in order that the memory of the past may not be blotted out from among men by time and that great and marvelous deeds done by Greeks and foreigners and especially the reason why they warred against each other may not lack renown" (Book 1.1)

📖 Read Herodotus' account of the Scythians

Write three paragraphs comparing the two civilizations.

- ❑ Paragraph one should summarize the Amazons and the Scyths; their dress, culture, practices, lifestyles, religion, etc.
- ❑ Paragraph two should retell how the two civilizations decided to merge.
- ❑ Paragraph three will tell the results of their merger and how it exists today. (This will require some additional research.) Also use a biblical principle to close the third paragraph.
- ❑ For each paragraph use six openers, two dress-ups, two decorations, and one triple extension.

L-18d: Warrior Goddesses

Bonus Women are often viewed as symbols of inspiration for the warrior. In many societies she also is a warrior and personifies victory.

1. How is this portrayed in the goddess legends?
   - ❑ Consider Greek Athena, Artemis, and Nike, Egyptian Isis, Minoan mother-goddess, Amazons, etc.
2. How does this compare to examples from the Bible?
   - ❑ Consider Deborah, Jael, Rahab, Abigail, etc.
3. Write an essay comparing the two views.
   - ❑ Use examples and excerpts to support the comparison.
   - ❑ Use all openers, two dress-ups, two decorations, and one triple extension with the appropriate markings.
4. Can you think of modern versions of women who are warriors or who personify victory? Write an additional paragraph including these modern examples and your opinion on their effectiveness to inspire. Who inspires you?

L-18e: Short Story – Completion

Finish your story and design a cover in an "ancient" style. Refer to **L-13e** for specifications.

Literature
Selections

Literature Table of Contents

The Story of Creation
Norse Myth

Of old there was nothing,
Nor sand, nor sea, nor cool waves.
No earth, no heaven above.
Only the yawning chasm
The sun knew not her dwelling.
Nor the moon his realm.
The stars had not their places.

Three roots there are to Yggdrasil
Hel lives beneath the first.
Beneath the second the frost-giants,
And men beneath the third.

The gods are doomed and the end is death.

The sun turns black, earth sinks in the sea,
The hot stars fall from the sky,
And fire leaps high about heaven itself.

In wondrous beauty once again.
The dwellings roofed with gold.
The fields unsowed bear ripened fruit
In happiness forevermore.

But I dare not ever to speak his name.
And there are few who can see beyond.
The moment when Odin falls.

The Story of Creation
Greek Myth

First there was Chaos, the vast immeasurable abyss,
Outrageous as a sea, dark, wasteful, wild.

Black winged Night
Into the bosom of Erebus dark and deep
Laid a wind-borne egg, and as the seasons rolled
Forth sprang Love, the longed for, shining, with
                    wings of gold.

Earth, the beautiful, rose up,
Broad-bosomed, she that is the steadfast base
Of all things And fair Earth first bore
The starry Heaven, equal to herself,
To cover her on all sides and to be
A home forever for the blessed gods.

A dreadful sound troubled the boundless sea.
The whole earth uttered a great cry.
Wide heaven, shaken, groaned.
From its foundation far Olympus reeled
Beneath the onrush of the deathless gods,
And trembling seized upon black Tartarus.

Bound in bitter chains beneath the wide-wayed earth,
As far below the earth as over earth
Is heaven, for even so far down lies Tartarus.
Nine days and nights would a bronze anvil fall
And on the tenth reach earth from heaven
And then again falling nine days and nights,
Would come to Tartarus, the brazen-faced.

To a high-piercing, headlong rock
In adamantine chains that none can break.

Forever shall the intolerable present grind you down.
And he who will release you is not born.
Such fruit you reap for your man-loving ways.
A god yourself, you did not dread God's anger,
But gave to mortals honor not their due.
And therefore you must guard this joyless rock—

No rest, no sleep, no moment's respite.
Groans shall your speech be, lamentation your only words.

Go and persuade the sea wave not to break.
You will persuade me no more easily.

An eagle red with blood
Shall come, a guest unbidden to your banquet.
All day long he will tear to rags your body,
Feasting in fury on the blackened liver.

There is no force that can compel my speech.
So let Zeus hurl his blazing bolts,
And with the white wings of the snow,
With thunder and with earthquake,
Confound the reeling world.
None of all this will bend my will.

Look for no ending to this agony
Until a god will freely suffer for you,
Will take on him your pain, and in your stead
Descend to where the sun is turned to darkness,
The black depth of death.

To bear on his back forever
The cruel strength of the crushing world
And the vault of the sky.
Upon his shoulders the great pillar
That holds apart the earth and heaven,
A load not easy to be borne.

A flaming monster with a hundred heads,
Who rose up against all the gods
Death whistled from his fearful jaws.
His eyes flashed glaring fire.

The bolt that never sleeps,
Thunder with breath of flame.
Into his very heart the fire burned.
His strength was turned to ashes.
And now he lies a useless thing
By Aetna, whence sometimes there burst
Rivers red-hot consuming with fierce jaws
The level fields of Sicily,
Lovely with fruits.
And that is Typhon's anger boiling up,
His fire-breathing darts.

Their boon is life forever freed from toil.
No more to trouble earth or the seawaters
With their strong hands,
Laboring for the food that does not satisfy.
But with the honored of the gods they live
A life where there are no more tears.
Around those blessed isles soft sea winds breathe.
And flowers of gold are blazing on the trees,
Upon the waters, too.

And now, though feeble and short-lived,
Mankind has flaming fire and there from
Learns many crafts.

Hurakan and Gucumatz
Central America
Story of Creation

At the beginning of time, all things were under the water except the givers of life, Hurakan and Gucumatz. They hovered over the water like giant hummingbirds. When the time came, Hurakan and Gucumatz said the word 'Earth', and slowly land was formed and rose up out of the water.

The creators saw that the land was flat, and caused hills and mountains to appear. Then they flew over the landscape that had been created, and the earth was at once covered with vegetation.

Next, Hurakan and Gucumatz made animals and set them down on the earth. They commanded the animals to worship them, but the animals could not understand or speak. They grunted and snuffled, roared and howled, but Hurakan and Gucumatz were not satisfied. They decided to make men more intelligent than the animals. These new people would know their creators and offer the praise and worship they wanted. Hurakan and Gucumatz debated what to do with the animals, and eventually decided to leave them on the earth so the men could kill them and eat them for food.

First, they modeled some human figures out of clay and placed them on the earth. But the clay men could not move or speak or understand, they were worse than the animals. Hurakan and Gucumatz could think of no way of making the clay men come to life, so they destroyed them.

Next, they carved some wooden men and stood them on the land. The wooden men looked perfect, and were much stronger than the clay men, being made out of living wood, they could move about, but they still had no intelligence of feelings and could not recognize their creators. Hurakan and Gucumatz decided to destroy them and start again, but this time some escaped and became the monkeys that we see today in the forests.

Hurakan and Gucumatz talked for a long time about the problem of making men who would be intelligent enough to worship their creators, but not so intelligent that they would want to overthrow the gods. Eventually they decided to make four men out of the life-giving plant, maize. They took strands of the yellow and white maize and carefully twisted and plaited them into the shapes of four men. This time, they were successful. The men they had made were perfect; they could run, jump, think, talk, eat and drink.

Indeed, in one way, they were too perfect; their eyesight was so good that they could see Hurakan and Gucumatz quite clearly, high up in the heavens. The gods held no mystery for them when they could be seen so easily, and Hurakan and Gucumatz realized that if they wanted the men to worship them, they would have do something about their eyesight. So they waited until the men were asleep, and then carefully shortened their sight so that they could no longer see into the heavens. Now the gods appeared mysterious to the men, and they began to worship their creators as Hurakan and Gucumatz desired.

The creator gods were so pleased with the worship of the four men that they made four women for them, to be their helpers and companions. So human life on earth had begun.

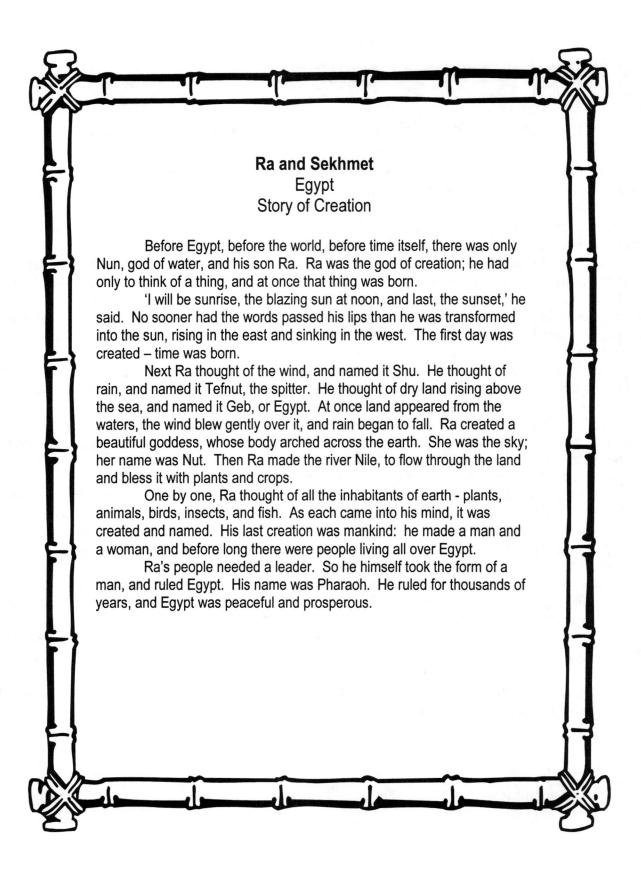

Ra and Sekhmet
Egypt
Story of Creation

Before Egypt, before the world, before time itself, there was only Nun, god of water, and his son Ra. Ra was the god of creation; he had only to think of a thing, and at once that thing was born.

'I will be sunrise, the blazing sun at noon, and last, the sunset,' he said. No sooner had the words passed his lips than he was transformed into the sun, rising in the east and sinking in the west. The first day was created – time was born.

Next Ra thought of the wind, and named it Shu. He thought of rain, and named it Tefnut, the spitter. He thought of dry land rising above the sea, and named it Geb, or Egypt. At once land appeared from the waters, the wind blew gently over it, and rain began to fall. Ra created a beautiful goddess, whose body arched across the earth. She was the sky; her name was Nut. Then Ra made the river Nile, to flow through the land and bless it with plants and crops.

One by one, Ra thought of all the inhabitants of earth - plants, animals, birds, insects, and fish. As each came into his mind, it was created and named. His last creation was mankind: he made a man and a woman, and before long there were people living all over Egypt.

Ra's people needed a leader. So he himself took the form of a man, and ruled Egypt. His name was Pharaoh. He ruled for thousands of years, and Egypt was peaceful and prosperous.

The Flood
by Ovid

The floods, by nature enemies to land,
And proudly swelling with their new command,
Remove the living stones that stopped their way,
And, gushing from their source, augment the sea.
Then, with his mace, their monarch struck the ground.
With inward trembling earth received the wound;
And rising streams a ready passage found.
The expanded waters gather on the plain,
They float the fields, and overtop the grain;
Then rushing onwards, with a sweepy sway,
Bear flocks, and folds, and laboring hinds away,
Nor safe their dwellings were; for, sapped by floods,
Their houses fell upon their household gods.
The solid piles, too strongly built to fall,
High o'er their heads behold a watery wall.
Now seas and earth were in confusion lost;
A world of waters, and without a coast.

One climbs a cliff; one in his boat is borne,
And ploughs above, where late he sowed his corn.
Others o'er chimney tops and turrets row,
And drop their anchors on the meads below;
Or downward driven, they bruise the tender vine,
Or tossed aloft, are knocked against a pine.
And where of late the kids had cropped the grass,
The monsters of the deep now take their place.
Insulting Nerieds on the cities ride,
And wondering dolphins o'er the palace glide.
On leaves, and masts of mighty oaks, they browse;
And their broad fins entangle in the boughs.
The frightened wolf now swims among the sheep;
The yellow lion wanders in the deep;
His rapid force no longer helps the boar;
The stag swims faster than he ran before.
The fowls, long beating on their wings in vain,
Despair of land, and drop into the main.
Now hills and vales no more distinction know,
And leveled nature lies oppressed below.
The most of mortals perish in the flood,
The small remainder dies for want of food.
A mountain of stupendous height there stands
betwixt the Athenian and Boeotian lands,
The bound of fruitful fields, while fields they were,

But then a field of waters did appear.
Parnassus is its name; whose forky rise
Mounts through the clouds, and mates the lofty skies.
High on the summit of this dubious cliff,
Deucalion wafting moored his little skiff.
He with his wife were only left behind
Of perished men; they two were human kind.
The mountain-nymphs and Themis they adore,
And from her oracles relief implore.
The most upright of mortal men was he:
The most sincere and holy woman, she.

When Jupiter, surveying earth from high,
Beheld it in a lake of water lie,
That, where so many millions lately lived,
But two, the best of either sex, survived,
He loosed the northern wind; fierce Boreas flies
To puff away the clouds, and purge the skies,
Serenely, while he blows, the vapors driven
Discover heaven to earth, and earth to heaven.
The billows fall, while Neptune lays his mace
On the rough sea, and smoothes its furrowed face,
Already Triton, at this call, appears
Above the waves; a Tyrian robe he wears;
And in his hand a crooked trumpet bears.
The sovereign bids him peaceful sounds inspire,
And give the waves the signal to retire.
His writhen shell he takes, whose narrow vent
Grows by degrees into a large extent;
Then gives it breathe; the blast, with doubling sound,
Runs the wide circuit of the world around.
The sun first heard it, in his early East,
And met the rattling echoes in the West.
The waters, listening to the trumpet's roar,
Obey the summons, and forsake the shore.

A thin circumference of land appears,
And Earth, but not at once, her visage rears,
And peeps upon the seas from upper grounds.
The streams, but just contained within their bounds,
By slow degrees into their channels crawl;
And earth increases as the waters fall.
In longer time the tops of trees appear,
Which mud on their dishonored branches bear.

At length the world was all restored to view,
But desolate, and of a sickly hue.

Nature beheld herself, and stood aghast,
A dismal desert, and a silent waste.

Which when Deucalion, with a piteous look,
Beheld, he wept, and thus to Pyrrha spoke:
"Oh wife, oh sister, oh of all thy kind
The best and only creature left behind,
By kindred, love, and now by dangers joined:
Of multitudes, who breathed the common air,
We two remain; a species in a pair;
The rest the seas have swallowed; nor have we
E'en of this wretched life a certainty.
The clouds are still above; and, while I speak,
A second deluge o'er our heads may break.
Should I be snatched from hence, and thou remain,
Without relief, or partner of thy pain,
How couldst thou such a wretched life sustain?
Should I be left, and thou be lost, the sea,
That buried her I loved, should bury me.
Oh could our father his old arts inspire,
And make me heir of his informing fire,
That so I might abolished man retrieve,
And perished people in new souls might live!
But heaven is pleased, nor ought we to complain,
That we, the examples of mankind remain."

He said: the careful couple join their tears,
And then invoke the gods, with pious prayers.
Thus in devotion having eased their grief,
From sacred oracles they seek relief;
And to Cephisus' brook their way pursue.
The stream was troubled, but the ford they knew.
With living waters in the fountain bred,
They sprinkle first their garments, and their head,
Then took the way which to the temple led.
The roofs were all defiled with moss and mire,
The desert altars void of solemn fire.
Before the gradual prostrate they adored,
The pavement kissed; and thus the saint implored:
"O righteous Themis, if the powers above
By prayers are bent to pity, and to love;
If human miseries can move their mind;
If yet they can forgive, and yet be kind;
Tell how we may restore, by second birth,
Mankind, and people desolated earth."

Then thus the gracious goddess, nodding, said:
"Depart, and with your vestments veil your head;
And stooping lowly down, with loosened zones,
Throw each behind your backs your mighty mother's bones."

Amazed the pair, and mute with wonder, stand,
Till Pyrrha first refused the dire command.
"Forbid it heaven," said she, "that I should tear
Those holy relics from the sepulcher."
They pondered the mysterious words again,
For some new sense; and long they sought in vain.
At length Deucalion cleared his cloudy brow,
And said: "The dark enigma will allow
A meaning, which, if well I understand,
From sacrilege will free the god's command.
This earth our mighty mother is; the stones
In her capacious body are her bones.
These we must cast behind, with hope, and fear,
The woman did the new solution hear.
The man confides in his own augury,
And doubts the gods; yet both resolve to try.

Descending from the mount, they first unbind
Their vests, and veiled, they cast the stones behind.
The stones (a miracle to mortal view,
But long tradition makes it pass for true)
Did first the rigor of their kind expel,
And supplied into softness as they fell;
Then swelled, and, swelling, by degrees grew warm;
And took the rudiments of human form;
Imperfect shades, in marble such are seen,
When the rude chisel does the man begin;
While yet the roughness of the stone remains,
Without the rising muscles, and the veins.
The sappy parts, and next resembling juice,
Were turned to moisture, for the body's use,
Supplying humors, blood, and nourishment.
The rest, too solid to receive a bent,
Converts to bones; and what was once a vein,
Its former name and nature did retain.
By help of power divine, in little space,
What the man threw, assumed a manly face;
And what the wife, renewed the female race.
Hence we derive our nature, born to bear
Laborious life, and hardened into care.

Ovid. "The Flood". Trans. Dryden, John

The Yellow Emperor and the Great Flood
China
Story of the Flood

High in the heavens, the Yellow Emperor sat musing on the wicked ways of mankind. Everywhere he looked on earth, he saw nothing but evil. Nothing, he decided, would bring man to his senses. Unless...a flood? Thought led at once to action. The Emperor of Heaven sent for Kung-kung, Spirit of Water, and gave him his instructions.

Kung-kung was a cruel god who took delight in using the power of water to cause suffering. Chuckling, he raced across the skies, causing torrential rain to fall on earth. For many days it rained, until the rivers burst their banks, and the floodwaters swirled through every town and village. Fearfully, men and women gathered up their belongings and made for higher ground. Many were drowned as the floodwaters overtook them, and those who escaped were soon starving for want of food.

Only one of the gods was moved to pity by the plight of mankind; Kun, grandson of the Yellow Emperor. He went to his grandfather and pleaded with him to recall the Spirit of Water and allow the floodwaters to abate. But the Yellow Emperor would not relent.

Sadly, Kun walked away from the imperial court, wondering what he could do to help mankind. As he walked, a gruff voice broke in on his thought.

'Why are you so miserable, Kun?' the voice inquired. It was a black tortoise, one of the oldest and wisest of creatures.

'I grieve for the suffering of mankind,' replied Kun, 'I would dearly like to save the world from the flood, but I know not how.'

The tortoise snorted. 'No difficulty about that,' he said, in his deep, slow voice. 'All you need is a handful of Magic Mould. Magic Mould is a special sort of earth; sprinkle it on water and it grows and expands to any amount you want. You'll only need a little.'

'But where would I get it from?' inquired Kun.

'From your grandfather, the Emperor, of course. Who else?'

Kun sighed. 'that's no good, then,' he said. 'He would never give me any.'

'So steal some!'

Kun sat down, and the two of them worked out a plan for stealing a handful of the Magic Mould. No one knows quite how Kun did it; but of course, being a god, he could change himself into many different shapes to deceive the guards and get into the room where the Magic Mould was kept.

When at last he had a supply of the Magic Mould, Kun set off to earth to try it out. Sitting on top of a high mountain, he sprinkled a few grains of the Mould onto the surface of the floodwaters beneath his feet. The crumbs of Mould sank beneath the water – and as Kun watched, the waters began to shrivel, as if they were being soaked up by a giant sponge. Soon what had been a lake became a waterlogged valley; gradually the marshy bottom dried out as well.

Kun was delighted. He walked down into the valley to inspect the results. Everywhere, the ground was firm and dry. From caves and treetops, people who had survived the flood began to appear. They looked down into the dry, brown valley, hardly able to believe their eyes. They scrambled down the mountainsides and knelt on the earth, running their fingers through the soil, laughing and crying in relief and delight. Some who had had the foresight to save some seeds began to plant them at once, and others set to work rebuilding their huts on the now firm and solid valley floor. Life had begun again.

Manu and the Fish
India
Story of the Flood

Manu was a wise and holy man who devoted his life to the worship of the gods. One day he was praying by the banks of a river when he heard a tiny voice calling.

'O great and holy Manu,' the voice said, "I appeal to you for help. You're a holy man; it's your duty to help and protect the weak."

Manu looked around to see who was speaking. But there was no one there. Then the voice came again. 'In here,' it said. Astonished, Manu saw that it was coming form a tiny fish in the river in front of him.

Carefully, he filled his cupped hands with water, bent and brought out the fish.

'Tell me what I can do for you, little one,' he said.

'Help me,' said the fish, 'You see how small I am. In the river I'm the smallest living thing; my life is in danger from enemies of every size.'

'What must I do to help?' asked Manu.

Put me in a jar, safe on land,' said the fish. 'In return, I'll save your life as well. A great flood is coming; all mankind will be swept away. If you help me, you alone will escape.'

Manu did not take this promise very seriously; even if there was flood, how could a tiny fish save him? Nevertheless, he fetched a large clay jar, filled it with water and placed the little fish in it.

Now this was no ordinary fish. In no time it had grown too big for the jar, and Manu had to dig a pond for it outside. Before long, even the pond was too small, and Manu carried the fish that was now huge, to the river Ganges.

Before it flopped into the river and made off to the ocean, the fish spoke again to Manu. It told him when the flood would come, and gave him instructions. He was to build a large boat, in good time, so that he would be safe when all the land was flooded. In it, he was to put seeds of every kind of plant, and a length of strong rope. The fish promised that he himself would reappear, and save Manu when the floods came. Manu would recognize him by a large horn on his head.

As the fish swam away, Manu realized that it was not really a fish at all, but the god Vishnu, preserver of life, himself. No mortal could look at a god and live; so Vishnu had appeared to him as a fish, in a form he could recognize and understand.

Respectfully, Manu bowed his head; when he looked up, the fish had gone. Manu hurried home and began to build a boat, as he had been instructed. He gathered seeds of every kind of plant, and plaited a length of strong rope. When all was done, he waited for the flood.

When the storms began and the waters rose, Manu set out in his boat across the sea. Before long, the great fish reappeared, and Manu made a noose in his rope and fastened it to the horn on the fish's head. Towed by the fish, Manu's boat was safe from all danger.

At last the waters began to subside. The boat grounded on a mountain peak high in the Himalayas, and the fish ordered Manu to moor it to the rock. Manu was the only survivor of the flood that destroyed mankind. He was the first of a new race, the father of all.

The Story of the Flood
from *The Epic of Gilgamesh*

You know the city Shurrupak, it stands on the banks of Euphrates? That city grew old and the gods that were in it were old. There was Anu, lord of the firmament, their father, and warrior Enlil their counselor, Ninurta the helper, and Ennugi watcher over canals; and with them also was Ea. In those days the world teemed, the people multiplied, the world bellowed like a wild bull, and the great god was aroused by the clamor. Enlil heard the clamor and he said to the gods in council, "The uproar of mankind is intolerable and sleep is no longer possible by reason of the babel." So the gods agreed to exterminate mankind. Enlil did this, but Ea, because of his oath, warned me in a dream. He whispered their words to my house of reeds, "Reed-house, reed-house! Wall, O wall, hearken reed-house, wall reflect: O man of Shurrupak, son of Ubara-Tutu: tear down your house and build a boat, abandon possessions and look for life, despise worldly goods and save your soul alive. Tear down your house, I say, and build a boat. These are the measurements of the boat as you shall build her: let her beam equal her length, let her deck be roofed like the vault that covers the abyss; then take up into the boat the seed of all living creatures."

'When I had understood I said to my lord, "Behold, what you have commanded I will honor and perform, but how shall I answer the people, the city, the elders?" Then Ea opened his mouth and said to me, his servant, "Tell them this: I have learned that Enlil is wrathful against me, I dare no longer walk in his land nor live in his city; I will go down to the Gulf to dwell with Ea my lord. But on you he will rain down abundance, rare fish and shy wildfowl, a rich harvest-tide. In the evening the rider of the storm will bring you wheat in torrents."

'In the first light of dawn all my household gathered round me, the children brought pitch and the men whatever was necessary. On the fifth day I laid the keel and the ribs, then I made fast the planking. The ground-space was one acre; each side of the deck measured one hundred and twenty cubits, making a square. I built six decks below, seven in all, I divided them into nine sections with bulkheads between. I drove in wedges where needed, I saw to the punt-poles, and laid in supplies. The carriers brought oil in baskets, I poured pitch into the furnace and asphalt and oil; more oil was consumed in caulking, and more again the master of the boat took into his stores. I slaughtered bullocks for the people and every day I killed sheep. I gave the shipwrights wine to drink as though it were river water, raw wine and red wine and oil and white wine. There was feasting then as there is at the time of the New Year's festival; I myself anointed my head. On the seventh day the boat was complete.

'Then was the launching full of difficulty; there was shifting of ballast above and below till two-thirds was submerged. I loaded into her all that I had of gold and of living things, my family, my kin, the beasts of the field both wild and tame, and all the craftsmen. I sent them on board, for the time that Shamash had ordained was already fulfilled when he said, "In the evening, when the rider of the storm sends down the destroying rain, enter the boat and batten her down." The time was fulfilled, the evening came, the rider of the storm sent down the rain. I looked out at the weather and it was terrible, so I too boarded the boat and battened her down. All was now complete, the battening and the caulking; so I handed the tiller to Puzur-Amurri the steersman, with the navigation and the care of the whole boat.

With the first light of dawn a black cloud came from the horizon; it thundered within where Adad, lord of the storm was riding. In front over hill and plain Shullat and Hanish, heralds of the storm, led on. Then the gods of the abyss rose up; Nergal pulled out the dams of the nether waters, Ninurta the war-lord threw down the dykes, and the seven judges of hell, the Annunake, raised their torches, lighting the land with their livid flame. A stupor of despair went up to heaven when the god of the storm turned daylight to darkness, when he smashed the land like a cup. One whole day the tempest raged, gathering fury as it went, it poured over the people like the tides of battle; a man could not see his brother nor the people be seen from heaven. Even the gods were terrified at the flood, they fled to the highest heaven, the firmament of Anu; they crouched against the walls, cowering like curs. Then Ishtar, the sweet-voiced Queen of Heaven cried out like a woman in travail: "Alas the days of old are turned to dust because I commanded evil; why did I command this evil in the council of all the gods? I commanded wars to destroy the people, but are they not my people, for I brought them forth? Now like the spawn of fish they float in the ocean." The great gods of heaven and of hell wept, they covered their mouths.

'For six days and six nights the winds blew, torrent and tempest and flood overwhelmed the world, tempest and flood raged together like warring hosts. When the seventh day dawned the storm from the south subsided, the sea grew calm, the flood was stilled; I looked at the face of the world and there was silence, all mankind was turned to clay. The surface of the sea stretched as flat as a rooftop; I opened a hatch and the light fell on my face. Then I bowed low, I sat down and I wept, the tears streamed down my face, for on every side was the waste of water. I looked for land in vain, but fourteen leagues distant there appeared a mountain, and there the boat grounded; on the mountain of Nisir the boat held fast, she held fast and did not budge. One day she held, and a second day on the mountain of Nisir she held fast and did not budge. A third day, and a fourth day she held fast on the mountain and did not budge; a fifth day and a sixth day she held fast on the mountain. When the seventh day dawned I loosed a dove and let her go. She flew away, but finding no resting-place she returned. Then I loosed a swallow, and she flew away but finding no resting-place she returned. I loosed a raven, she saw that the waters had retreated, she ate, she flew around, she cawed, and she did not come back. Then I threw everything open to the four winds, I made a sacrifice and poured out a libation on the mountaintop. Seven and again seven cauldrons I set up on their stands, I heaped up wood and cane and cedar and myrtle. When the gods smelled the sweet savor, they gathered like flies over the sacrifice. Then, at last, Ishtar also came. She lifted her necklace with the jewels of heaven that once Anu had made to please her. "O you gods here present, by the lapis lazuli round my neck I shall remember these days as I remember the jewels of my throat; these last days I shall not forget. Let all the gods gather round the sacrifice, except Enlil. He shall not approach this offering, for without reflection he brought the flood; he consigned my people to destruction."

'When Enlil had come, when he saw the boat, he was wroth and swelled with anger at the gods, the host of heaven, "Has any of these mortals escaped? Not one was to have survived the destruction." Then the god of the wells and canals Ninurta opened his mouth and said to the warrior Enlil, "Who is there of the gods that can devise without Ea? It is Ea alone who knows all things." Then Ea opened his mouth and spoke to warrior Enlil, "Wisest of gods, hero Enlil, how could you so senselessly bring down the flood?

> *Lay upon the sinner his sin,*
> *Lay upon the transgressor his transgression,*
> *Punish him a little when he breaks loose,*
> *Do not drive him too hard or he perishes;*
> *Would that a lion had ravaged mankind*
> *Rather than the flood,*
> *Would a wolf had ravaged mankind*
> *Rather than the flood,*
> *Would that famine had wasted the world*
> *Rather than the flood,*
> *Would that pestilence had wasted the mankind*
> *Rather than the flood.*

It was not I that revealed the secret of the gods; the wise man learned it in a dream. Now take your counsel what shall be done with him."

'Then Enlil went into the boat, he took me by the hand and my wife and made us enter the boat and kneel down on either side, he standing between us. He touched our foreheads to bless us saying, "In time past Utnapishtim was a mortal man; henceforth he and his wife shall live in the distance at the mouth of the rivers." Thus it was that the gods took me and placed me here to live in the distance, at the mouth of the rivers.'

The Story of the Flood. "The Epic of Gilgamesh." New York: Penguin Classics, 1972. Pgs. 108 -113

The Great Flood
Deucalion's Flood

When evil first came among mankind, people became very wicked. War, robbery, treachery, and murder prevailed throughout the world. Even the worship of the gods, the laws of truth and honor, reverence for parents and brotherly love were neglected.

Finally, Zeus determined to destroy the race of men altogether, and the other gods agreed. All the winds were therefore shut up in a cave except the South Wind, the wet one. He raced over the earth with water streaming from his beard and long, white hair. Clouds gathered around his head, and dew dripped from his wings and the ends of his garments. With him went Iris, the rainbow goddess, while below Poseidon smote the earth with his trident until it shook and gaped open, so that the waters of the sea rushed up over the land.

Fields and farmhouses were buried. Fish swam in the tops of the trees. Sea beasts were quietly feeding where flocks and herds had grazed before. On the surface of the water, boars, stags, lions, and tigers struggled desperately to keep afloat. Wolves swam in the midst of flocks of sheep, but they did not frighten the sheep, and the wolves never thought of their natural prey. Each fought for his life and forgot the other. Over them wheeled countless birds, winging far and wide in the hope of finding something to rest upon. Eventually they fell into the water and were drowned.

All over the water were men in small boats or makeshift rafts. Some even had oars that they tried to use, but the waters were fierce and stormy, and there was nowhere to go. In time all were drowned, until at last there was no one left but an old man and his wife, Deucalion and Pyrrha. These two people had lived in truth and justice, unlike the rest of mankind. They had been warned of the coming of the flood and had built a boat and stocked it. For nine days and nights they floated until Zeus took pity on them and they came to the top of Mount Parnassus, the sacred home of the Muses. There they found land and disembarked to wait while the gods recalled the water they had unloosed.

When the waters fell, Deucalion and Pyrrha looked over the land, despairing. Mud and sea slime covered the earth; all living things had been swept away. Slowly and sadly they made their way down the mountain until they came to a temple where there had been an oracle. Black seaweed dripped from the pillars now, and the mud was over all. Nevertheless the two knelt down and kissed the temple steps while Deucalion prayed to the goddess to tell them what they should do. All men were dead but themselves, and they were old. It was impossible that they should have children to people the earth again. Out of the temple a great voice was heard speaking strange words.

"Depart," it said, "with veiled heads and loosened robes, and throw behind you as you go the bones of your mother."

Pyrrha was in despair when she heard this saying. "The bones of our mother!" she cried. "How can we tell now where they lie? Even if we knew, we could never do such a dreadful thing as to disturb their resting place and scatter them over the earth like an armful of stones."

"Stones!" said Deucalion quickly. "That must be what the goddess means. After all Earth is our mother, and the other thing is too horrible for us to suppose that a goddess would ever command it."

Accordingly both picked up armfuls of stones, and as they went away from the temple with faces veiled, they cast the stones behind them. From each of those Deucalion cast sprang up a man, and from Pyrrha's stones sprang women. Thus the earth was re-peopled, and in the course of time it brought forth again animals from itself, and all was as before. Only from that time men have been less sensitive and have found it easier to endure toil, and sorrow, and pain, since now they are descended from stones.

Coolidge, Olivia. "The Great Flood". Greek Myths. Boston, Houghton Co., 1949. pgs 73 - 75

The Coming of Evil
The Story of Pandora
Greek Myth

    After the punishment of Prometheus, Zeus planned to take his revenge on man. He could not recall the gift of fire, since it had been given by one of the immortals, but he was not content that man should possess this treasure in peace and become perhaps as great as were the gods themselves. He therefore took counsel with the other gods, and together they made for man a woman. All the gods gave gifts to this new creation. Aphrodite gave her fresh beauty like the spring itself. The goddess Athena dressed her and put on her a garland of flowers and green leaves. She had also a golden diadem beautifully decorated with figures of animals. In her heart Hermes put cunning, deceit, and curiosity. She was named Pandora, which means All-Gifted, since each of the gods had given her something. The last gift was a chest in which there was supposed to be great treasure, but which Pandora was instructed never to open. Then Hermes, the Messenger, took the girl and brought her to Epimetheus.

    Epimetheus had been warned by his brother to receive no gifts from Zeus, but he was a heedless person, as ever, and Pandora was very lovely. He accepted her, therefore, and for awhile they lived together in happiness, for Pandora, besides her beauty, had been given both wit and charm. Eventually, however, her curiosity got the better of her, and she determined to see for herself what treasure it was that the gods had given her. One day, when she was alone, she went over to the corner where her chest lay and cautiously lifted the lid for a peep. The lid flew up out of her hands and knocked her aside, while before her frightened eyes dreadful, shadowy shapes flew out of the box in an endless stream. There were hunger, disease, war, greed, anger, jealousy, toil, and all the griefs and hardships to which man from that day has been subject. Each was terrible in appearance, and as it passed, Pandora saw something of the misery that her thoughtless action had brought on her descendants. At last the stream slackened, and Pandora, who had been paralyzed with fear and horror, found strength to shut her box. The only thing left in it now, however, was the one good gift the gods had put in among so many evil ones. This was hope, and since that time the hope that is in man's heart is the only thing that has made him able to bear the sorrows that Pandora brought upon him.

Coolidge, Olivia. "The Coming of Evil". Greek Myths, Boston: Houghton Mifflen Co., 1949, pgs 70 - 72

Summary of Part 1
The Epic of Gilgamesh

⌐ ⌐ ⌐ ⌐ ⌐ ⌐ ⌐ ⌐ ⌐ ⌐ ⌐ ⌐ ⌐ ⌐ ⌐

The epic begins by introducing us to Gilgamesh, strong, handsome, and brave, who loved the city Uruk that he had built. Yet the citizens of Uruk complained bitterly to the great god Anu of the arrogance, ruthlessness, and depravity of Gilgamesh. When Anu heard their cries he directed the goddess Aruru saying, "You made him, O Aruru, now create his equal." So Aruru created Enkidu, a huge, strong, hairy creature who lived among the wild beasts. In the steppes Enkidu frees wild beasts from traps and fills up holes dug by trappers, causing them great frustration. Gilgamesh in turn sets a trap for Enkidu. A prostitute was sent to seduce and civilize Enkidu. She is successful in fulfilling her mission. Coming into Uruk, Enkidu challenges Gilgamesh for lording it over the people. A fight ensued ending in peace and mutual affection.

Part 2 The Forest Journey

Enlil of the mountain, the father of the gods, had decreed the destiny of Gilgamesh. So Gilgamesh dreamed and Enkidu said, 'The meaning of the dream is this. The father of the gods has given you kingship, such is your destiny, everlasting life is not your destiny. Because of this do not be sad at heart, do not be grieved or oppressed. He has given you power to bind and to loose, to be the darkness and the light of mankind. He has given you unexampled supremacy over the people, victory in battle from which no fugitive returns, in forays and assaults from which there is no going back. But do not abuse this power, deal justly with your servants in the palace, deal justly before Shamash.'

The eyes of Enkidu were full of tears and his heart was sick. He sighed bitterly and Gilgamesh met his eye and said, 'My friend, why do you sigh so bitterly?' But Enkidu opened his mouth and said, 'I am weak, my arms have lost their strength, the cry of sorrow sticks in my throat, I am oppressed by idleness.' It was then that the lord Gilgamesh turned his thoughts to the Country of the Living; on the Land of Cedars and the lord Gilgamesh reflected. He said to his servant Enkidu, "I have not established my name stamped on bricks as my destiny decreed; therefore I will go to the country where the cedar is felled. I will set up my name in the place where the names of famous men are written, and where no man's name is written yet I will raise a monument to the gods. Because of the evil that is in the land, we will go to the forest and destroy the evil; for in the forest lives Humbaba whose name is "Hugeness", a ferocious giant.' But Enkidu sighed bitterly and said, 'When I went with the wild beasts ranging through the wilderness I discovered the forest; its length is ten thousand leagues in every direction. Enlil has appointed Humbaba to guard it and armed him in sevenfold terrors, terrible to all flesh is Humbaba. When he roars it is like the torrent of the storm, his breath is like fire, and his jaws are death itself. He guards the cedars so well that when the wild heifer stirs in the forest, though she is sixty leagues distant, he hears her. What man would willingly walk into that country and explore its depths? I tell you, weakness overpowers whoever goes near it: it is not an equal struggle when one fights Humbaba; he is a great warrior, a battering-ram. Gilgamesh, the watchman of the forest never sleeps.'

Gilgamesh replied: 'Where is the man who can clamber to heaven? Only the gods live forever with glorious Shamash, but as for us men, our days are numbered, our occupations are a breath of wind. How is this, already you are afraid! I will go first although I am your lord, and you may safely call out, "Forward, there is nothing to fear!" Then if I fall I leave behind me a name that endures; men will say of me, "Gilgamesh has fallen in fight with ferocious Humbaba." Long after the child has been born in my house, they will say it, and remember.' Enkidu spoke again to Gilgamesh, 'O my lord, if you will enter that country, go first to the hero Shamash, tell the Sun God, for the land is his. The country where the cedar is cut belongs to Shamash.'

Gilgamesh took up a kid, white without spot, and a brown one with it; he held them against his breast, and he carried them into the presence of the sun. He took in his hand his silver scepter and he said to glorious Shamash, 'I am going to that country, O Shamash, I am going; my hands supplicate, so let it be well with my soul and bring me back to the quay of Uruk. Grant, I beseech, your protection, and let the omen be good.' Glorious Shamash answered, 'Gilgamesh, you are strong, but what is the Country of the Living to you?'

'O Shamash, hear me, hear me, Shamash, let my voice be heard. Here in the city man dies oppressed at heart, man perishes with despair in his heart. I have looked over the wall and I see the bodies floating on the river, and that will be my lot also. Indeed I know it is so, for whoever is tallest among men cannot reach the heavens and the greatest cannot encompass the earth. Therefore I would enter that country: because I have not established my name stamped on brick as my destiny decreed, I will go to the country where the cedar is cut. I will set up my name where the names of famous men are written; and where no man's name is written I will raise a monument to the gods.' The tears ran down his face and he said, 'Alas, it is a long journey that I must take to the Land of Humbaba. If this enterprise is not to be accomplished, why did you move me, Shamash, with the restless desire to perform it? How can I succeed if you will not succor me? If I die in that country I will die without rancor, but if I return I will make a glorious offering of gifts and of praise to Shamash.'

So Shamash accepted the sacrifice of his tears; like the compassionate man he showed him mercy. He appointed strong allies for Gilgamesh, sons of one mother, and stationed them in the mountain caves. The great winds he appointed; the north wind, the whirlwind, the storm and the icy wind, the tempest and the scorching wind. Like vipers, like dragons, like a scorching fire, like a serpent that freezes the heart, a destroying flood and the lightning's fork, such were they, and Gilgamesh rejoiced.

He went to the forge and said, 'I will give orders to the armourers; they shall cast us our weapons while we watch them.' So they gave orders to the armourers and the craftsmen sat down in conference. They went into the groves of the plain and cut willow and box wood; they cast for them axes of nine score pounds, and great swords they cast with blades of six score pounds each one, with pommels and hilts of thirty pounds. They cast for Gilgamesh the axe 'Might of Heroes' and the bow of Anshan; and Gilgamesh was armed and Enkidu; and the weight of the arms they carried was thirty score pounds.

The people collected and the counselors in the streets and in the marketplace of Uruk; they came through the gate of seven bolts and Gilgamesh spoke to them in the marketplace: 'I, Gilgamesh, go to see that creature of whom such things are spoken, the rumor of whose name fills the world. I will conquer him in his cedar wood and show the strength of the sons of Uruk, all the world shall know of it. I am committed to this enterprise: to climb the mountain, to cut down the cedar, and leave behind me an enduring name.' The counselors of Uruk, the great market, answered him, 'Gilgamesh, you are young, your courage carries you too far, you cannot know what this enterprise means which you plan. We have heard that Humbaba is not like men who die, his weapons are such that none can stand against them; the forest stretches for ten thousand leagues in every direction; who would willingly go down to explore its depths? As for Humbaba, when he roars it is like the torrent of the storm, his breath is like fire and his jaws are death itself. Why do you crave to do this thing, Gilgamesh? It is no equal struggle when one fights with Humbaba, that battering-ram.'

When he heard these words of the counselors Gilgamesh looked at his friend and laughed, 'How shall I answer them; shall I say I am afraid of Humbaba, I will sit at home all the rest of my days? Then Gilgamesh opened his mouth again and said to Enkidu, 'My friend, let us go to the Great Palace, to Egalmah, and stand before Ninsun the queen. Ninsun is wise with deep knowledge, she will give us counsel for the road we must go.' They took each other by the hand as they went to Egalmah, and they went to Ninsun the great queen. Gilgamesh approached, he entered the palace and spoke to Ninsun. 'Ninsun, will you listen to me; I have a long journey to go, to the Land of Humbaba, I must travel an unknown road and fight a strange battle. From the day I go until I return, till I reach the cedar forest and destroy the evil which Shamash abhors, pray for me to Shamash.'

Ninsun went into her room, she put on a dress becoming to her body, she put on jewels to make her breast beautiful, she placed a tiara on her head and her skirts swept the ground. Then she went to the altar of the Sun, standing upon the roof of the palace; she burnt incense and lifted her arms to Shamash as the smoke ascended: 'O Shamash, why did you give this restless heart to Gilgamesh, my son; why did you give it? You have moved him and now he sets out on a long journey to the Land of Humbaba, to travel an unknown road and fight a strange battle. Therefore from the day that he goes till the day he returns, until he reaches the cedar forest, until he kills Humbaba and destroys the evil thing which you, Shamash, abhor, do not forget him; but let the dawn, Aya, your dear bride, remind you always and when day is done give him to the watchman of the night to keep him from harm.'

Then Ninsun the mother of Gilgamesh extinguished the incense, and she called to Enkidu with this exhortation: 'Strong Enkidu, you are not the child of my body, but I will receive you as my adopted son; you are my other child like the foundlings they bring to the temple. Serve Gilgamesh as a foundling serves the temple and the priestess who reared him. In the presence of my women, my votaries and hierophants, I declare it.' Then she placed the amulet for a pledge round his neck, and she said to him, 'I entrust my son to you; bring him back to me safely.'

And now they brought to them the weapons, they put in their hands the great swords in their golden scabbards, and the bow and the quiver. Gilgamesh took the axe, he slung the quiver from his shoulder, and the bow of Anshan, and buckled the sword to his belt; and so they were armed and ready for the journey. Now all the people came and pressed on them and said, 'When will you return to the city?' The counselors blessed Gilgamesh and warned him, 'Do not trust too much in your own strength, be watchful, restrain your blows at first. The one who goes in front protects his companion; the good guide who knows the way guards his friend. Let Enkidu lead the way, he knows the road to the forest, he has seen Humbaba and is experienced in battles; let him press first into the passes, let him be watchful and look to himself. Let Enkidu protect his friend, and guard his companion, and bring him safe through the pitfalls of the road. We, the counselors of Uruk entrust our king to you, O Enkidu; bring him back safely to us.' Again to Gilgamesh they said, 'May Shamash give you your heart's desire, may he let you see with your eyes the thing accomplished which your lips have spoken; may he open a path for you where it is blocked, and a road for your feet to tread. May he open the mountains for your crossing, and may the nighttime bring you the blessings of night, and Lugulbanda, your guardian god, stand beside you for victory. May you have victory in the battle as though you fought with a child. Wash your feet in the river of Humbaba to which you are journeying; in the evening dig a well, and let there always be pure water in your water-skin. Offer cold water to Shamash and do not forget Lugulbanda.'

Then Enkidu opened his mouth and said, 'Forward, there is nothing to fear. Follow me, for I know the place where Humbaba lives and the paths where he walks. Let the counselors go back. Here is no cause for fear.' When the counselors heard this they sped the hero on his way. 'Go, Gilgamesh, may your guardian god protect you on the road and bring you safely back to the quay of Uruk.'

After twenty leagues they broke their fast; after another thirty leagues they stopped for the night. Fifty leagues they walked in one day; in three days they had walked as much as a journey of a month and two weeks. They crossed seven mountains before they came to the gate of the forest. Then Enkidu called out to Gilgamesh, 'Do not go down into the forest; when I opened the gate my hand lost its strength.' Gilgamesh answered him, 'Dear friend, do not speak like a coward. Have we got the better of so many dangers and traveled so far, to turn back at last? You, who are tired in wars and battles, hold close to me now and you will feel no fear of death; keep beside me and your weakness will pass, the trembling will leave your hand. Would my friend rather stay behind? No, we will go down together into the heart of the forest. Let your courage be roused by the battle to come; forget death and follow me, a man resolute in action, but one who is not foolhardy. When two go together each will protect himself and shield his companion, and if they fall they leave an enduring name.'

Together they went down into the forest and they came to the green mountain. There they stood still, they were struck dumb; they stood still and gazed at the forest. They saw the height of the cedar, they saw the way into the forest and the track where Humbaba used to walk. The way was broad and the going was good. They gazed at the mountain of cedars, the dwelling place of the gods, and throne of Ishtar. The hugeness of the cedar rose in front of the mountain, its shade was beautiful, full of comfort; mountain and glade were green with brushwood.

There Gilgamesh dug a well before the setting sun. He went up the mountain and poured out fine meal on the ground and said, 'O mountain, dwelling of the gods, bring me a favorable dream.' Then they took each other by the hand and lay down to sleep; and sleep that flows from the night lapped over them. Gilgamesh dreamed, and at midnight sleep left him, and he told his dream to his friend. 'Enkidu, what was it that woke me if you did not? My friend, I have dreamed a dream. Get up, look at the mountain precipice. The sleep that the gods sent me is broken. Ah, my friend, what a dream I have had! Terror and confusion; I seized hold of a wild bull in the wilderness. It bellowed and beat up the dust till the whole sky was dark, my arm was seized and my tongue bitten. I fell back on by knee; then someone refreshed me with water from his water-skin.'

Enkidu said, 'Dear friend, the god to whom we are traveling is no wild bull, though his form is mysterious. That wild bull which you saw is Shamash the Protector; in our moment of peril he will take our hands. The one who gave water from his water skin, that is your own god who cares for your good name, your Lugulbanda. United with him, together we will accomplish a work the fame of which will never die.'

Gilgamesh said, 'I dreamed again. We stood in a deep gorge of the mountain, and beside it we two were like the smallest of swamp flies; and suddenly the mountain fell, it struck me and caught my feet from under me. Then came an intolerable light blazing out, and in it was one whose grace and whose beauty were greater than the beauty of this world. He pulled me out from under the mountain, he gave me water to drink and my heart was comforted, and he set my feet on the ground.'

Then Enkidu the child of the plains said, 'Let us go down from the mountain and talk this thing over together.' He said to Gilgamesh the young god, 'Your dream is good, your dream is excellent, the mountain which you saw is Humbaba. Now, surely, we will seize and kill him, and throw his body down as the mountain fell on the plain.'

The next day after twenty leagues they broke their fast, and after another thirty they stopped for the night. They dug a well before the sun had set and Gilgamesh ascended the mountain. He poured out fine meal on the ground and said, 'O mountain, dwelling of the gods, send a dream for Enkidu, make him a favorable dream.' The mountain fashioned a dream for Enkidu; it came, an ominous dream; a cold shower passed over him, it caused him to cower like the mountain barley under a storm of rain. But Gilgamesh sat with his chin on his knees till the sleep which flows over all mankind lapped over him. Then, at midnight, sleep left him; he got up and said to his friend, 'Did you call me, or why did I wake? Did you touch me, or why am I terrified? Did not some god pass by, for my limbs are numb with fear? My friend, I saw a third dream and this dream was altogether frightful. The heavens roared and the earth roared again, daylight failed and darkness fell, lightnings flashed, fire blazed out, the clouds lowered, they rained down death. Then the brightness departed, the fire went out, and all was turned to ashes fallen about us. Let us go down from the mountain and talk this over, and consider what we should do.'

When they had come down from the mountain Gilgamesh seized the axe in his hand: he felled the cedar. When Humbaba heard the noise far off he was enraged; he cried out, 'Who is this that has violated my woods and cut down my cedar?' But glorious Shamash called to them out of heaven, 'Go forward, do not be afraid.' But now Gilgamesh was overcome by weakness, for sleep had seized him suddenly, a profound sleep held him; he lay on the ground, stretched out speechless, as though in a dream. When Enkidu touched him he did not rise, when he spoke to him he did not reply. "O Gilgamesh, Lord of the plain of Kullab, the world grows dark, the shadows have spread over it, now is the glimmer of dusk. Shamash has departed, his bright head is quenched in the bosom of his mother Ningal. O Gilgamesh, how long will you lie like this, asleep? Never let the mother who gave you birth be forced in mourning into the city square.'

At length Gilgamesh heard him; he put on his breastplate, 'The Voice of Heroes, of thirty shekels' weight; he put it on as though it had been a light garment that he carried, and it covered him altogether. He straddled the earth like a bull that snuffs the ground and his teeth were clenched. 'By the life of my mother Ninsun who gave me birth, and by the life of my father, divine Lugulbanda, let me live to be the wonder of my mother, as when she nursed me on her lap.' A second time he said to him, 'By the life of my father, divine Lugulbanda, until we have fought this man, if man he is, this god, if god he is, the way that I took to the Country of the Living will not turn back to the city.'

Then Enkidu, the faithful companion, pleaded, answering him, 'O my lord, you do not know this monster and that is the reason you are not afraid. I who know him, I am terrified. His teeth are dragon's fangs, his countenance is like a lion, his charge is the rushing of the flood, with his look he crushes alike the trees of the forest and reeds in the swamp. O my Lord, you may go on if you choose into this land, but I will go back to the city. I will tell the lady your mother all your glorious deeds till she shouts for joy: and then I will tell the death that followed till she weeps for bitterness.' But Gilgamesh said, 'Immolation and sacrifice are not yet for me, the boat of the dead shall not go down, nor the three ply cloth be cut for my shrouding. Not yet will my people be desolate, nor the pyre be lit in my house and my dwelling burnt on the fire.

Today, give me your aid and you shall have mine: what then can go amiss with us two? All living creatures born of the flesh shall sit at last in the boat of the West, and when it sinks, when the boat of Magilum sinks, they are gone; but we shall go forward and fix our eyes on this monster. If your heart is fearful throw away fear; if there is terror in it throw away terror. Take your axe in your hand and attack. He who leaves the fight unfinished is not at peace.'

Humbaba came out from his strong house of cedar. Then Enkidu called out, 'O Gilgamesh, remember now your boasts in Uruk. Forward, attack, son of Uruk, there is nothing to fear.' When he heard these words his courage rallied; he answered, 'Make haste, close in, if the watchman is there do not let him escape to the woods where he will vanish. He has put on the first of his seven splendors but not yet the other six, let us trap him before he is armed.' Like a raging wild bull he snuffed the ground; the watchman of the woods turned full of threatening, he cried out. Humbaba came from his strong house of cedar. He nodded his head and shook it, menacing Gilgamesh; and on him he fastened his eye, the eye of death. Then Gilgamesh called to Shamash and his tears were flowing, 'O glorious Shamash, I have followed the road you commanded but now if you send no succor how shall I escape?' Glorious Shamash heard his prayer and he summoned the great wind, the north wind, the whirlwind, the storm and the icy wind, the tempest and the scorching wind; they came like dragons, like a scorching fire, like a serpent that freezes the heart, a destroying flood and the lightning's fork. The eight winds rose up against Humbaba, they beat against his eyes; he was gripped, unable to go forward or back. Gilgamesh shouted, 'By the life of Ninsun my mother and divine Lugulbanda my father, in the Country of the Living, in this Land I have discovered your dwelling; my weak arms and my small weapons I have brought to this Land against you, and now I will enter your house'.

So he felled the first cedar and they cut the branches and laid them at the foot of the mountain. At the first stroke Humbaba blazed out, but still they advanced. They felled seven cedars and cut and bound the branches and laid them at the foot of the mountain, and seven times Humbaba loosed his glory on them. As the seventh blaze died out they reached his lair. He slapped his thigh in scorn. He approached like a noble wild bull roped on the mountain, a warrior whose elbows are bound together. The tears started to his eyes and he was pale, 'Gilgamesh, let me speak. I have never known a mother, no nor a father who reared me. I was born of the mountain, he reared me, and Enlil made me the keeper of this forest. Let me go free, Gilgamesh, and I will be your servant, you shall be my lord; all the trees of the forest that I tended on the mountain shall be yours. I will cut them down and build you a palace.' He took him by the hand and led him to his house, so that the heart of Gilgamesh was moved with compassion. He swore by the heavenly life, by the earthly life, by the underworld itself: 'O Enkidu, should not the snared bird return to it nest and the captive man return to his mother's arms?' Enkidu answered, 'The strongest of men will fall to fate if he has no judgment. Namtar, the evil fate that knows no distinction between men, will devour him. If the snared bird returns to its nest, if the captive man returns to his mother's arms, then you my friend will never return to the city where the mother is waiting who gave you birth. He will bar the mountain road against you, and make the pathways impassable.'

Humbaba said, 'Enkidu, what you have spoken is evil: you, a hireling, dependent for your bread! In envy and for fear of a rival you have spoken evil words.' Enkidu said, 'Do not listen, Gilgamesh: this Humbaba must die. Kill Humbaba first and his servants after.' But Gilgamesh said, 'If we touch him the blaze and the glory of light will be put out in confusion, the glory and glamour will vanish, its rays will be quenched.' Enkidu said to Gilgamesh, 'Not so, my friend. First entrap the bird, and where shall the chicks run then? Afterwards we can search out the glory and the glamour, when the chicks run distracted through the grass.'

Gilgamesh listened to the word of his companion, he took the axe in his hand, he drew the sword from his belt, and he struck Humbaba with a thrust of the sword to the neck, and Enkidu his comrade struck the second blow. At the third blow Humbaba fell. Then there followed confusion for this was the guardian of the forest whom they had felled to the ground. For as far as two leagues the cedars shivered when Enkidu felled the watcher of the forest, he at whose voice Hermon and Lebanon used to tremble. Now the mountains were moved and all the hills, for the guardian of the forest was killed. They attacked the cedars, the seven splendors of Humbaba were extinguished. So they pressed on into the forest bearing the sword of eight talents. They uncovered the sacred dwellings of the Anunnaki and while Gilgamesh felled the first of the trees of the forest Enkidu cleared their roots as far as the banks of Euphrates. They set Humbaba before the god, before Enlil; they kissed the ground and dropped the shroud and set the head before him. When he saw the head of Humbaba, Enlil raged at them. 'Why did you do this thing? From henceforth may the fire be on your faces, may it eat the bread that you eat, may it drink where you drink.' Then Enlil took again the blaze and the seven splendors that had been Humbaba's: he gave the first to the river, and he gave to the lion, to the stone of execration, to the mountain and to the dreaded daughter of the Queen of Hell.

O Gilgamesh, king and conqueror of the dreadful blaze; wild bull who plunders the mountain, who crosses the sea, glory to him, and from the brave the greater glory is Enki's!

Part 3 Ishtar and Gilgamesh, and the Death of Enkidu

Gilgamesh washed out his long locks and cleaned his weapons; he flung back his hair from his shoulders; he threw off his stained clothes and changed them for new. He put on his royal robes and made them fast. When Gilgamesh had put on the crown, glorious Ishtar lifted her eyes, seeing the beauty of Gilgamesh. She said, 'Come to me Gilgamesh, and be my bridegroom; grant me seed of your body, let me be your bride and you shall be my husband. I will harness for you a chariot of lapis lazuli and of gold, with wheels of gold and horns of copper; and you shall have mighty demons of the storm for draft mules. When you enter our house in the fragrance of cedar wood, threshold and throne will kiss your feet. Kings, rulers, and princes will bow down before you; they shall bring you tribute from the mountains and the plain. Your ewes shall drop twins and your goats triplets; your pack ass shall outrun mules; your oxen shall have no rivals, and your chariot horses shall be famous far-off for their swiftness.'

Gilgamesh opened his mouth and answered glorious Ishtar, 'If I take you in marriage, what gifts can I give in return? What ointments and clothing for your body? I would gladly give you bread and all sorts of food fit for a god. I would give you wine to drink fit for a queen. I would pour out barley to stuff your granary; but as for making you my wife - that I will not. How would it go with me? Your lovers have found you like a brazier which smolders in the cold, a back door which keeps out neither squall of wind nor storm, a castle which crushes the garrison, pitch that blackens the bearer, a water skin that chafes the carrier, a stone which falls from the parapet, a battering ram turned back from the enemy, a sandal that trips the wearer. Which of your lovers did you ever love forever? What shepherd of yours has pleased you for all time? Listen to me while I tell the tale of your lovers. There was Tammuz, the lover of your youth, for him you decreed wailing, year after year. You loved the many-colored roller, but still you struck and broke his wing; now in the grove he sits and cries, "kappi, kappi, my wing, my wing." You have loved the lion tremendous in strength: seven pits you dug for him, and seven. You have loved the stallion magnificent in battle, and for him you decreed whip and spur and a thong, to gallop seven leagues by force and to muddy the water before he drinks; and for his mother Silili lamentations. You have loved the shepherd of the flock; he made meal cake for you day after day, he killed kids for your sake. You struck and turned him into a wolf; now his own herd boys chase him away, his own hounds worry his flanks. And did you not love Ishullanu, the gardener of your father's palm grove? He brought you baskets filled with dates without end; every day he loaded your table. Then you turned your eyes on him and said, "Dearest Ishullanu, come here to me, let us enjoy your manhood, come forward and take me, I am yours." Ishullanu answered, "What are you asking from me? My mother has baked and I have eaten; why should I come to such as you for food that is tainted and rotten? For when was a screen of rushes sufficient protection from frosts?" But when you had heard his answer you struck him. He was changed to a blind mole deep in the earth, one whose desire is always beyond his reach. And if you and I should be lovers, should not I be served in the same fashion as all these others whom you loved once?'

When Ishtar heard this she fell into a bitter rage, she went up to high heaven. Her tears poured down in front of her father Anu, and Antum her mother. She said, 'My father, Gilgamesh has heaped insults on me, he has told over all my abominable behavior, my foul and hideous acts.' Anu opened his mouth and said, "Are you a father of gods"? Did not you quarrel with Gilgamesh the king, so now he has related your abominable behavior, your foul and hideous acts.'

Ishtar opened her mouth and said again, 'My father, give me the Bull of Heaven to destroy Gilgamesh. Fill Gilgamesh, I say, with arrogance too his destruction; but if you refuse to give me the Bull of Heaven I will break in the doors of hell and smash the holds; there will be confusion of people, those above with those from the lower depths. I shall bring up the dead to eat food like the living; and the hosts of dead will outnumber the living.' Anu said to great Ishtar, 'If I do what you desire there will be seven years of drought throughout Uruk when corn will be seedless husks. Have you saved grain enough for the people and grass for the cattle?' Ishtar replied. "I have saved grain for the people, grass for the cattle; for seven years of seedless husks there is grain and there is grass enough.'

When Anu heard what Ishtar had said he gave her the Bull of Heaven to lead by the halter down to Uruk. When they reached the gates of Uruk the Bull went to the river; with his first snort cracks opened in the earth and a hundred young men fell down to death. With his second snort cracks opened and two hundred fell down to death. With his third snort cracks opened, Enkidu doubled over but instantly recovered, he dodged aside and leapt on the Bull and seized it by the horns. The Bull of Heaven foamed in his face, it brushed him with the thick of its tail. Enkidu cried to Gilgamesh, 'My friend, we boasted that we would leave enduring names behind us. Now thrust in your sword between the nape and the horns.' So Gilgamesh followed the Bull, he seized the thick of its tail, he thrust the sword between the nape and the horns and slew the Bull. When they had killed the Bull of Heaven. They cut out its heart and gave it to Shamash, and the brothers rested.

But Ishtar rose up and mounted the great wall of Uruk; she sprang on to the tower and uttered a curse: 'Woe to Gilgamesh, for he has scorned me in killing the Bull of Heaven.'   When Enkidu heard these words he tore out the Bull's right thigh and tossed it in her face saying, 'If I could lay my hands on you, it is this I should do to you, and lash the entrails to your side.' Then Ishtar called together her people, the dancing and singing girls, the prostitutes of the temple, the courtesans. Over the thigh of the Bull of Heaven she set up lamentation.

But Gilgamesh called the smiths and the armourers, all of them together. They admired the immensity of the horns. They were plated with lapis lazuli two fingers thick. They were thirty pounds each in weight, and their capacity in oil was six measures, which he gave to his guardian god, Lugulbanda. But he carried the horns into the palace and hung them on the wall. Then they washed their hands in Euphrates, they embraced each other and went away. They drove through the streets of Uruk where the heroes were gathered to see them, and Gilgamesh called to the singing girls, "Who is most glorious of the heroes, who is most eminent among men?" Gilgamesh is the most glorious of heroes, Gilgamesh is most eminent among men.' And now there was feasting, and celebrations and joy in the palace, till the heroes lay down saying, 'Now we will rest for the night.'

When the daylight came Enkidu got up and cried to Gilgamesh, 'O my brother, such a dream I had last night. Anu, Enlil, Ea and heavenly Shamash took counsel together, and Anu said to Enlil, "Because they have killed the Bull of Heaven, and because they have killed Humbaba who guarded the Cedar Mountain one of the two must die." Then glorious Shamash answered the hero Enlil, "It was by your command they killed the Bull of Heaven, and killed Humbaba, and must Enkidu die although innocent?" Enlil flung round in rage at glorious Shamash, "You dare to say this, you who went about with them every day like one of themselves!"

So Enkidu lay stretched out before Gilgamesh; his tears ran down in streams and he said to Gilgamesh, "O my brother, so dear as you are to me, brother, yet they will take me from you.' Again he said, 'I must sit down on the threshold of the dead and never again will I see my dear brother with my eyes.'

While Enkidu lay alone in his sickness he cursed the gate as though it was living flesh, 'You there, wood of the gate, dull and insensible, witless, I searched for you over twenty leagues until I saw the towering cedar. There is no wood like you in our land. Seventy-two cubits high and twenty-four wide, the pivot and the ferrule and the jambs are perfect. A master craftsman from Nippur has made you; but O, if I had known the conclusion! If I had known that this was all the good that would come of it, I would have raised the axe and split you into little pieces and set up here a gate of wattle instead. Ah, if only some future king had brought you here, or some god had fashioned you. Let him obliterate my name and write his own, and the curse fall on him instead of on Enkidu.'

With the first brightening of dawn Enkidu raised his head and wept before the Sun God, in the Brilliance of the sunlight his tears streamed down. "Sun God, I beseech you, about the vile Trapper, that Trapper of nothing because of whom I was to catch less than my comrade; let him catch least, make his game scarce, make him feeble, taking the smaller of every share, let his quarry escape from his nets.'

When he had cursed the Trapper to his heart's content he turned on the harlot. He was roused to curse her also. 'As for you, woman, with a great curse I curse you! I will promise you a destiny to all eternity. My curse shall come on you soon and sudden. You shall be without a roof for your commerce, for you shall not keep house with other girls in the tavern, but do your business in places fouled by the vomit of the drunkard. Your hire will be potter's earth, your thieving will be flung into the hovel, you will sit at the crossroads in the dust of the potter's quarter, you will make your bed on the dunghill at night, and by day take your stand in the wall's shadow. Brambles and thorns will tear your feet, the drunk and the dry will strike your cheek and your mouth will ache. Let you be stripped of your purple dyes, for I, too, once in the wilderness with my wife had all the reassure I wished.'

When Shamash heard the words of Enkidu he called to him from heaven: 'Enkidu, why are you cursing the woman, the mistress who taught you to eat bread fit for gods and drink wine of kings? She who put upon you a magnificent garment, did she not give you glorious Gilgamesh for your companion, and has not Gilgamesh, your own brother, made you rest on a royal bed and recline on a couch at his left hand? He has made the princes of the earth kiss your feet, and now all the people of Uruk lament and wail over you. When you are dead he will let his hair grow long for your sake, he will wear a lion's pelt and wander through the desert.'

When Enkidu heard glorious Shamash his angry heart grew quiet, he called back the curse and said, 'Woman, I promise you another destiny. The mouth which cursed you shall bless you! Kings, princes and nobles shall adore you. On your account a man though twelve miles off will clap his hand to his thigh and his hair will twitch. For you he will undo his belt and open his treasure and you shall have your desire; lapis lazuli, gold and carnelian from the heap in the treasury. A ring for your hand and a robe shall be yours. The priest will lead you into the presence of the gods. On your account a wife, a mother of seven, was forsaken.'

As Enkidu slept alone in his sickness, in bitterness of spirit he poured out his heart to his friend. 'It was I who cut down the cedar, I who leveled the forest, I who slew Humbaba and now see what has become of me. Listen, my friend, this is the dream I dreamed last night. The heavens roared, and earth trembled back an answer; between them stood I before an awful being, the somber faced man-bird; he had directed on me his purpose. His was a vampire face, his foot was a lion's foot, his hand was an eagle's talon. He fell on me and his claws were in my hair, he held me fast and I smothered; then he transformed me so that my arms became wings covered with feathers. He turned his stare towards me, and he led me away to the palace of Irkalla, the Queen of Darkness, to the house from which none who enters ever returns, down the road from which there is no coming back.

'There is the house whose people sit in darkness; dust is their food and clay their meat. They are clothed like birds with wings for covering, they see no light, they sit in darkness. I entered the house of dust and I saw the kings of the earth, their crowns put away forever; rulers and princes, all those who once wore kingly crowns and ruled the world in the days of old. They who had stood in the place of the gods like Anu and Enlil, stood now like servants to fetch baked meats in the house of dust, to carry cooked meat and cold water from the water skin. In the house of dust which I entered were high priests and acolytes, priests of the incantation and of ecstasy; there were servers of the temple, and there was Etana, the king of Kish whom the eagle carried to heaven in the days of old. I saw also Samuqan, god of cattle, and there was Ereshkigal the Queen of the Underworld; and Belit-Sheri squatted in front of her, she who is recorder of the gods and keeps the book of death. She held a tablet from which she read. She raised her head, she saw me and spoke: "Who has brought this one here?" Then I awoke like a man drained of blood, who wanders alone in a waster of rushes; like one whom the bailiff has seized and his heart pounds with terror.'

Gilgamesh had peeled off his clothes, he listened to his words and wept quick tears, Gilgamesh listened and his tears flowed. He opened his mouth and spoke to Enkidu: 'Who is there in strong walled Uruk who has wisdom like this? Strange things have been spoken, why does your heart speak strangely? The dream was marvelous but the terror was great; we must treasure the dream whatever the terror; for the dream has shown that misery comes at last to the healthy man, the end of life is sorrow.' And Gilgamesh laments, 'Now I will pray to the great gods, for my friend had an ominous dream."

This day on which Enkidu dreamed came to an end and he lay stricken with sickness. One whole day he lay on his bed and his suffering increased. He said to Gilgamesh, the friend on whose account he had left the wilderness, "Once I ran for you, for the water of life, and I now have nothing." A second day he lay on his bed and

Gilgamesh watched over him but the sickness increased. A third day he lay on his bed, he called out to Gilgamesh, rousing him up. Now he was weak and his eyes were blind with weeping. Ten days he lay and his suffering increased, eleven and twelve days he lay on his bed of pain. The he called to Gilgamesh, 'My friend, the great goddess cursed me and I must die in shame. I shall not die like a man fallen in battle; I feared to fall, but happy is the man who falls in the battle, for I must die in shame.' And Gilgamesh wept over Enkidu. With the first light of dawn he raised his voice and said to the counselors of Uruk:

'Hear me, great ones of Uruk
I weep for Enkidu, my friend,
Bitterly moaning like a woman mourning.
I weep for my brother.
O Enkidu, my brother,
You were the axe at my side,
My hand's strength, the sword in my belt,
The shield before me,
A glorious robe, my fairest ornament;
An evil Fate has robbed me.
The wild ass and the gazelle
That were father and mother,
All long-tailed creatures that nourished you
Weep for you,
All the wild things of the plains and pastures;
The paths that you loved in the forest of cedars
Night and day murmur.
Let the great ones of strong walled Uruk
Weep for you;
Let the finger of blessing
Be stretched out in the mourning;
Enkidu, young brother, Hark,
There is an echo through all the country
Like a mother mourning.
Weep all the paths where we walked together;
And the beasts we hunted, the bear and hyena,
Tiger and panther, leopard and lion,
The stag and the ibex, the bull and the doe.
The river along whose banks we used to walk,
Weeps for you,
Ula of Elam and dear Euphrates
Where once we drew water for the water skins.
The mountain we climbed where we slew the Watchman,
Weeps for you.
The Warriors of strong walled Uruk
Where the Bull of Heaven was killed,
Weep for you.
All the people of Eridu
Weep for you Enkidu.
Those who brought grain for your eating
Mourn for you now;

Who rubbed oil on you back
Mourn for you now;
Who poured beer for your drinking
Mourn for you now.
The harlot who anointed you with fragrant ointment
Laments for you now;
The women of the palace, who brought you a wife,
A chosen ring of good advice,
Lament for you now.
And the young men your brothers
As though they were women
Go long-haired in mourning.
What is this sleep which holds you now?
You are lost in the dark and cannot hear me.'

He touched his heart but it did not beat, nor did he lift his eyes again. When Gilgamesh touched his heart it did not beat. So Gilgamesh laid a veil, as one veils the bride, over his friend. He began to rage like a lion, like a lioness robbed of her whelps. This way and that he paced round the bed, he tore out his hair and strewed it around. He dragged off his splendid robes and flung them down as though they were abominations.

In the first light of dawn Gilgamesh cried out, 'I made you rest on a royal bed, you reclined on a couch at my left hand, the princes of the earth kissed your feet. I will cause all the people Uruk to weep over you and raise the dirge of the dead. The joyful people will stoop with sorrow; and when you have gone to the earth I will let my hair grow long for your sake, I will wander through the wilderness in the skin of a lion.' The next day also, in the first light, Gilgamesh lamented' seven days and seven nights he wept for Enkidu, until the worm fastened on him. Only then he gave him up to the earth, for the Anunnaki, the judges, had seized him.

Then Gilgamesh issued a proclamation through the land, he summoned them all, the coppersmiths, the goldsmiths, the stone-workers, and commanded them, 'Make a statue of my friend.' The statue was fashioned with a great weight of lapis lazuli for the breast and of gold for the body. A table of hardwood was set out, and on it a bowl carnelian filled with honey, and a bowl of lapis lazuli filled with butter. These he exposed and offered to the Sun; and weeping he went away.

Part 4 The Search for Everlasting Life

Bitterly Gilgamesh wept for his friend Enkidu; he wandered over the wilderness as a hunter, he roamed over the plains; in his bitterness he cried, 'How can I rest, how can I be at peace? Despair is in my heart. What my brother is now, that shall I be when I am dead. Because I am afraid of death I will go as best I can to find Utnapishtim whom they call the Faraway, for he has entered the assembly of the gods.' So Gilgamesh traveled over the wilderness, he wandered over the grasslands, a long journey, in search of Utnapishtim, whom the gods took after the deluge; and they set him to live in the land of Dilmun, in the garden of the sun; and to him alone of men they gave everlasting life.

At night when he came to the mountain passes Gilgamesh prayed: 'In these mountain passes long ago I saw lions, I was afraid and I lifted my eyes to the moon; I prayed and my prayers went up to the gods, so now, O moon god Sin, protect me.' When he had prayed he lay down to sleep, until he was woken from out of a dream. He saw the lions round him glorying in life; then he took his axe in his hand, he drew his sword from his belt, and he fell upon them like an arrow from the string, and struck and destroyed and scattered them.

So at length Gilgamesh came to Mashu, the great mountains about which he had heard many things, which guard the rising and the setting sun. Its twin peaks are as high as the wall of heaven and its paps reach down to the underworld. At its gate the Scorpions stand guard, half man and half dragon; their glory is terrifying, their stare strikes death into men, their shimmering halo sweeps the mountains that guard the rising sun. When Gilgamesh saw them he shielded his eyes for the length of a moment only; then he took courage and approached. When they saw him so undismayed the Man-Scorpion called to his mate, 'This one who comes to us now is flesh of the gods.' The mate of the Man-Scorpion answered, 'Two thirds is god but one third is man.'

Then he called to the man Gilgamesh, he called to the child of the gods: 'Why have you come so great a journey; for what have you traveled so far, crossing the dangerous waters; tell me the reason for your coming?' Gilgamesh answered, 'For Enkidu; I loved him dearly, together we endured all kinds of hardships; on his account I have come, for the common lot of man has taken him. I have wept for him day and night, I would not give up his body for burial, I thought my friend would come back because of my weeping. Since he went, my life is nothing; that is why I have traveled here in search of Utnapishtim my father; for men say he has entered the assembly of the gods, and has found everlasting life. I have a desire to question him concerning the living and the dead.' The Man-Scorpion opened his mouth and said, speaking to Gilgamesh, 'No man born of woman has done what you have asked, no mortal man has gone into the mountain; the length of it is twelve leagues of darkness; in it there is no light, but the heart is oppressed with darkness. From the rising of the sun to the setting of the sun there is no light.' Gilgamesh said, 'Although I should go in sorrow and in pain, with sighing and with weeping, still I must go. Open the gate of the mountain.' And the Man-Scorpion said, 'Go, Gilgamesh, I permit you to pass through the mountain of Mashu and through the high ranges; may your feet carry you safely home. The gate of the mountain is open.'

When Gilgamesh heard this he did as the Man-Scorpion had said, he followed the sun's road to his rising, through the mountain. When he had gone one league the darkness became thick around him, for there was no light, he could see nothing ahead and nothing behind him. After two leagues the darkness was thick and there was no light, he could see nothing ahead and nothing behind him. After three leagues the darkness was thick, and there was no light, he could see nothing ahead and nothing behind him. After four leagues the darkness was thick and there was no light, he could see nothing ahead and nothing behind him. At the end of five leagues the darkness was thick and there was no light, he could see nothing ahead and nothing behind him. At the end of seven leagues the darkness was thick and there was no light, he could see nothing ahead and nothing behind him. When he had gone eight leagues Gilgamesh gave a great cry, for the darkness was thick and he could see nothing ahead and nothing behind him. After nine leagues he felt the north wind on his face, but the darkness was thick and there was no light, he could see nothing ahead and nothing behind him. After ten leagues the end was near. After eleven leagues the dawn light appeared. At the end of twelve leagues the sun streamed out.

There was the garden of the gods; all round him stood bushes bearing gems. Seeing it he went down at once, for there was fruit of carnelian with the vine hanging from it, beautiful to look at; lapis lazuli leaves hung thick with fruit, sweet to see. For thorns and thistles there were haematite and rare stones, agate, and pearls from out of the sea. While Gilgamesh walked in the garden by the edge of the sea Shamash saw him, and he saw that he was dressed in the skins of animals and ate their flesh. He was distressed, and he spoke and said, 'No mortal man has gone this way before, nor will, as long as the winds drive over the sea.' And to Gilgamesh he said, 'You will never find the life for which you are searching.' Gilgamesh said to glorious Shamash, 'Now that I have toiled and strayed so far over the wilderness, am I to sleep, and let the earth cover my head for ever: Let my eyes see the sun until they are dazzled with looking. Although I am no better than a dead man, still let me see the light of the sun.'

Beside the sea she lives, the woman of the vine, the maker of wine; Siduri sits in the garden at the edge of the sea, with the golden bowl and the golden vats that the gods gave her. She is covered with a veil; and where she sites she sees Gilgamesh coming towards her, wearing skins, the flesh of the gods in his body, but despair in his heart, and his face like the face of one who has made a long journey. She looked, and as she scanned the distance she said in her own heart, 'Surely this is some felon; where is he going now?' And she barred her gate against him with the crossbar and shot home the bolt. But Gilgamesh, hearing the sound of the bolt, threw up his head and lodged his foot in the gate; he called to her, 'Young woman, maker of wine, why do you bolt your door; what did you see that made you bar your gate? I will break in your door and burst in your gate, for I am Gilgamesh who seized and killed the Bull of Heaven, I killed the watchman of the cedar forest, I overthrew Humbaba who lived in the forest, and I killed the lions in the passes of the mountain.'

Then Siduri said to him, 'If you are that Gilgamesh who seized and killed the Bull of Heaven, who killed the watchman of the cedar forest, who overthrew Humbaba that lived in the forest, and killed the lions in the passes of the mountain, why are your cheeks so starved and why is your face so drawn? Why is despair in your heart and your face like the face of one who has made a long journey? Yes, why is your face burned from heat and cold, and why do you come here wandering over the pastures in search of the wind?

Gilgamesh answered her, 'And why should not my cheeks be starved and my face drawn? Despair is in my heart and my face is the face of one who has made a long journey, it was burned with heat and with cold. Why should I not wander over the pastures in search of the wind? My friend, my younger brother, he who hunted the wild ass of the wilderness and the panther of the plains, my friend, my younger brother who seized and killed the Bull of Heaven and overthrew Humbaba in the cedar forest, my friend who was very dear to me and who endures dangers beside me, Enkidu my brother, whom I loved, the end of mortality has overtaken him. I wept for him seven days and nights till the worm fastened on him. Because of my brother I am afraid of death, because of my brother I stray through the wilderness and cannot rest. But now, young woman, maker of wine, since I have seen your face do not let me see the face of death which I dread so much.'

She answered, 'Gilgamesh, where are you hurrying to? You will never find that life for which you are looking. When the gods created man they allotted to him death, but life they retained in their own keeping. As for you, Gilgamesh, fill your belly with good things; day and night, night and day, dance and be merry, feast and rejoice. Let your clothes be fresh, bathe yourself in water, cherish the little child that holds your hand, and make your wife happy in your embrace; for this too is the lot of man.'

But Gilgamesh said to Siduri, the young woman, 'How can I be silent, how can I rest, when Enkidu whom I love is dust, and I too shall die and be laid in the earth. You live by the seashore and look into the heart of it; young woman, tell me now, which is the way to Utnapishtim, the son of Ubara-Tutu? What directions are there for the passage; give me, oh, give me directions. I will cross the Ocean if it is possible; if it is not I will wander still farther in the wilderness.' The winemaker said to him, 'Gilgamesh, there is no crossing the Ocean; whoever has come, since the days of old, has not been able to pass that sea. The Sun in his glory crosses the Ocean, but who beside Shamash has ever crossed it? The place and the passage are difficult, and the waters of death are deep which flow between. Gilgamesh, how will you cross the Ocean? When you come to the waters of death what will you do? But Gilgamesh, down in the woods you will find Urshanbi, the ferryman of Utnapishtim; with him are the hold things, the things of stone. He is fashioning the serpent prow of the boat. Look at him well, and if it is possible, perhaps you will cross the waters with him; but if it is not possible, then you must go back.'

When Gilgamesh heard this he was seized with anger. He took his axe in his hand, and his dagger from his belt. He crept forward and he fell on them like a javelin. Then he went into the forest and sat down. Urshanabi saw the dagger flash and heard the axe, and he beat his head, for Gilgamesh had shattered the tackle of the boat in his rage. Urshanabi said to him, 'Tell me, what is your name, I am Urshanabi, the ferryman of Utnapishtim the Faraway.' He replied to him, 'Gilgamesh is my name, I am from Uruk, from the house of Anu.' Then Urshanabi said to him, 'Why are your cheeks so starved and your face drawn: Why is despair in your heart and your face like the face of one who has made a long journey; yes, why is your face burned with heat and with cold, and why do you come here wandering over the pastures in search of the wind?'

Gilgamesh said to him, 'Why should not my cheeks be starved and my face drawn? Despair is in my heart, and my face is the face of one who has made a long journey. I was burned with heat and with cold. Why should I not wander over the pastures: My friend, my younger brother who seized and killed the Bull of Heaven, and overthrew Humbaba in the cedar forest, my friend who was very dear to me, and who endured dangers beside me, Enkidu my brother whom I loved, the end of mortality has overtaken him. Because of my brother I am afraid of death, because of my brother I stray through the wilderness. His fate lies heavy upon me. How can I be silent, how can I rest? He is dust and I too shall die and be laid in the earth forever. I am afraid of death, therefore, Urshanabi, tell me which is the road to Utnapishtim? If it is possible I will cross the waters of death; if not I will wander still farther through the wilderness.'

Urshanabi said to him, 'Gilgamesh, your own hands have prevented you from crossing the Ocean; when you destroyed the tackle of the boat you destroyed its safety.' Then the two of them talked it over and Gilgamesh said, 'Why are you so angry with me, Urshanabi, for you yourself cross the sea by day and night, at all seasons you cross it.' 'Gilgamesh, those things you destroyed, their property is to carry me over the water, to prevent the waters of death from touching me. It was for this reason that I preserved them, but you have destroyed them. And the *urnu* snakes with them. But now, go into the forest, Gilgamesh; with your axe cut poles, one hundred and twenty, cut them sixty cubits long, paint them with bitumen, set on them ferrules and bring them back.'

When Gilgamesh heard this he went into the forest, he cut poles one hundred and twenty. He cut them sixty cubits long, he painted them with bitumen, he set on them ferrules, and he brought them to Urshanabi. Then they boarded the boat, Gilgamesh and Urshanabi together, launching it out on the waves of Ocean. For three days they ran on as it were a journey of a month and fifteen days, and at last Urshanabi brought the boat to the waters of death. Then Urshanabi said to Gilgamesh, 'Press on, take a pole and thrust it in, but do not let your hands touch the waters. Gilgamesh, take a second pole, take a third, take a fourth pole. Now, take fifth, sixth, and a seventh pole. Now, Gilgamesh, take an eighth, a ninth, a tenth pole. Gilgamesh, take an eleventh, take a twelfth pole.' After one hundred and twenty thrusts Gilgamesh had used the last pole. Then he stripped himself, he held up his arms for a mast and his covering for a sail. So Urshanabi the ferryman brought Gilgamesh to Utnapishtim, whom they call the Faraway, who lives in Dilmun at the place of the sun's transit, eastward of the mountain. To him alone of men the gods had given everlasting life.

Now Utnapishtim, where he lay at ease, looked into the distance and he said to his heart, musing to himself, 'Why does the boat sail here without tackle and mast; why are the sacred stones destroyed, and why does the master not sail the boat? That man who comes is none of mine; where I look I see a man whose body is covered with skins of beasts. Who is this who walks up the shore behind Urshanabi, for surely he is no man of mine?' So Utnapishtim looked at him and said, 'What is your name, you who come here wearing the skins of beasts, with your cheeks starved and your face drawn? Where are you hurrying to now? For what reason have you made this great journey, crossing the seas whose passage is difficult? Tell me the reason for your coming.'

He replied, 'Gilgamesh is my name. I am from Uruk, from the house of Anu.' Then Utnapishtim said to him, 'If you are Gilgamesh, why are your cheeks so starved and your face drawn? Why is despair in your heart and your face like the face of one who has made a long journey? Yes, why is your face burned with heat and cold; and why do you come here, wandering over the wilderness in search of the wind?'

Gilgamesh said to him, 'Why should not my cheeks be starved and my face drawn? Despair is in my heart and my face is the face of one who has made a long journey. It was burned with heat and with cold. Why should I not wander over the pastures? My friend, my younger brother who seized and killed the Bull of Heaven and overthrew Humbaba in the cedar forest, my friend who was very dear to me and endured dangers beside me, Enkidu, my brother whom I loved, the end of mortality has overtaken him. I wept for him seven days and nights till the worm fastened on him. Because of my brother I am afraid of death; because of my brother I stray through the wilderness. His fate lies heavy upon me. How can I be silent, how can I rest? He is dust and I shall die also and be laid in the earth forever.' Again Gilgamesh said, speaking to Utnapishtim, 'It is to see Utnapishtim whom we call the Faraway that I have come on this journey. For this I have wandered over the world, I have crossed many difficult ranges, I have crossed the seas, I have wearied myself with traveling; my joints are aching, and I have lost acquaintance with sleep that is sweet. My clothes were worn out before I came to the house of Siduri. I have killed the bear and hyena, the lion and panther, the tiger, the stag and the ibex, all sorts of wild game and the small creatures of the pastures. I ate their flesh and I wore their skins; and that was how I came to the gate of the young woman, the maker of wine, who barred her gate of pitch and bitumen against me. But from her I had news of the journey; so then I came to Urshanabi the ferryman, and with him I crossed over the waters of death. Oh, father Utnapishtim, you who have entered the assembly of the gods, I wish to question you concerning the living and the dead, how shall I find the life for which I am searching?'

Utnapishtim said, 'There is no permanence. Do we build a house to stand forever; do we seal a contract to hold for all time? Do brothers divide an inheritance to keep forever, does the flood time of rivers endure? It is only the nymph of the dragonfly who sheds her larva and sees the sun in his glory. From the days of old there is no permanence. The sleeping and the dead, how alike they are, they are like a painted death. What is there between the master and the servant when both have fulfilled their doom? When the Anunnaki, the judges, come together, and Mammetun the mother of destinies, together they decree the fates of men. Life and death they allot but the day of earth they do not disclose.'

Then Gilgamesh said to Utnapishtim the Faraway, 'I look at you now, Utnapishtim, and your appearance is no different from mine; there is nothing strange in your features. I thought I should find you like a hero prepared for battle, but you lie here taking your ease on your back. Tell me truly, how was it that you came to enter the company of the gods and to possess everlasting life?' Utnapishtim said to Gilgamesh, "I will reveal to you a mystery, I will tell you a secret of the gods.'

The Epic of Gilgamesh dates from the third millennium B.C. translated by N.K. Sandars.

Utnapishtim: He is the protégé of the god Ea, by whose connivance he survives the flood, with his family and with 'the seed of all living creatures'; afterwards the gods take him to live forever where the sun rises.

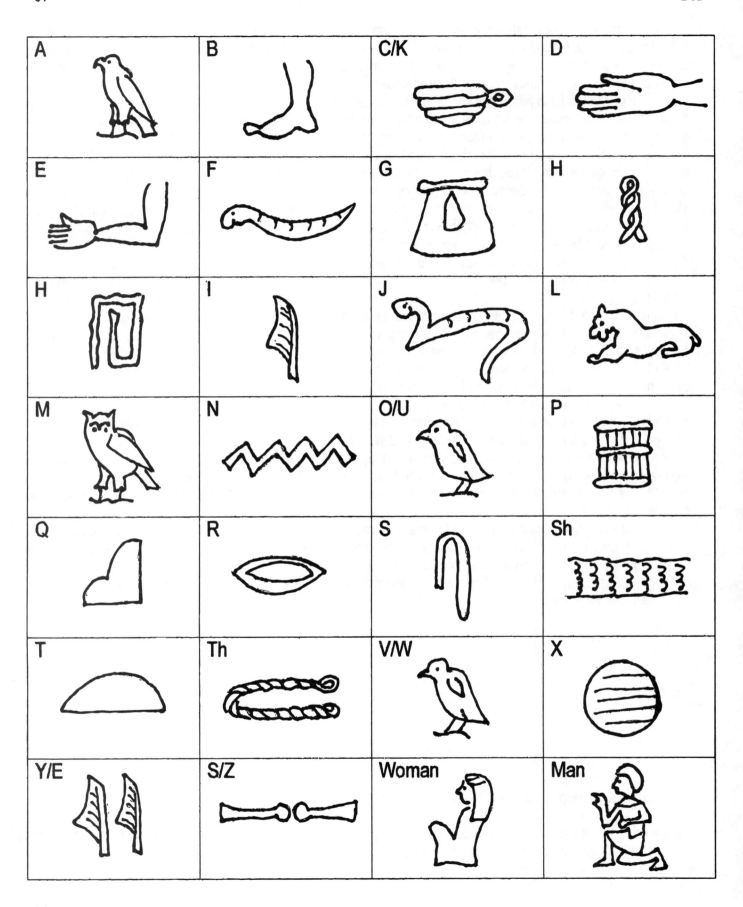

A	B	C/K	D
E	F	G	H
H	I	J	L
M	N	O/U	P
Q	R	S	Sh
T	Th	V/W	X
Y/E	S/Z	Woman	Man

Egyptian Poetry

I. HE:

I sail the "King's Waters,"
And then into the waterways of Heliopolis:
My destination the place of tents
At the entrance to Mertu Harbor.
I must hurry!
Restless, excited,
My heart goes out in prayer,
Prayer to the sun god Ra
For a quick, safe voyage.
I will be able to see her
As she walks beside the river.

SHE:

With you here at Mertu
Is like being at Heliopolis already
We return to the tree-filled garden,
My arms full of flowers.

Looking at my reflection in the still pool—
My arms full of flowers—
I see you creeping on tip-toe
To kiss me from behind,
My hair heavy with perfume.

With your arms around me
I feel as if I belong to the Pharaoh.

II.

The shrill of the wild goose
Unable to resist
The temptation of my bait.

While I, in a tangle of love,
Unable to break free,
Must watch the bird carry away my nets.
And when my mother returns, loaded with birds,
And finds me empty-handed,
What shall I say?

That I caught no birds?
That I myself was caught in your net?

The Story of Sinuhe
Egyptian Novel

Introduction

The hereditary prince and commander, warden and district officer of the estates of the sovereign in the lands of the Asiatics, this truly beloved royal agent, the follower Sinuhe, said:

I was a follower who followed his lord, a servant of the king's harem and of the hereditary princess, greatest of praise, wife of King Sesostris in Khnumet-sut and daughter of King Ammenemes in Ka-nofru, the possessor of an honored state.

Death of Ammenemes I and Sinuhe's Flight

Year 30, month 3 of Akhet, day 7. The god ascended to his horizon, the King of Upper and Lower Egypt, Sehetepibre. He flew up to the heavens, being joined to the sun disk, the god's body being mixed with that of him who made him.

The capital was silent, desires were weak, the Great Double Gate was locked, the court was with head upon knee, and nobles were wailing. Now His Majesty had dispatched an expeditionary force to the land of the Tjemehi with his eldest son as its leader, the good god, Sesostris. He had been sent to strike the foreign lands and to smite those who were among the Tjehenu people. And now he was returning, having brought back captives of the Tjehenu people and all kinds of cattle without number. The Companions of the Palace sent to the Western Half to inform the king's son of the affairs that had taken place in the council chamber. When the messengers found him upon the road and reached him at dusk, he did not delay for a moment. The Falcon flew off with his followers without letting his expeditionary force know it.

Now (they had) written to the sons of the king who were in his following in this expeditionary force. When it was being read out to one of them, I was standing by and I heard his voice, as he spoke, being in the vicinity not far away. My senses were disturbed, my arms spread out, and trembling came over every part (of me). I took myself off by stealth to find for myself a place of concealment. I placed myself between two shrubs in order to separate the road from its traveler. I went south. I did not plan to reach the capital, for I thought riots might occur, and I would not be able to say "life" after him (the king). I crossed the (place called) The Two Truths in the vicinity of the Sycamore, and I landed at The Island of Snefru. I spent the day on the edge of the cultivation. At dawn I came upon a man standing in the middle of the road. He greeted me respectfully, for he was frightened. When the time of evening meal came, I reached the wharf of Negau. It was in a boat without a rudder that I crossed over with the help of the west wind. I passed by to the east of the Quarry, above the Mistress of the Red Mountain (that is, the opposite side of the river, beyond the cultivation, in the desert), and I gave a path to my feet northward (until) I touched the Walls of the Prince, which had been made to check the Asiatics and to crush the sand-travelers. I took a crouched position in the brush out of fear that the guard on duty on the walls might see. I went by night, and when day came, I had reached Peten. I alighted at the Island of Kem-wer. Thirst overcame me; I was parched, my throat dry. And I said: This is the taste of Death. But I raised up my heart and gathered together my limbs. I heard the sound of the lowing of cattle, and I looked upon Asiatics. Their Bedouin chief perceived me, a man who had been in Egypt. He gave me water and boiled milk for me, and I went with him to his tribe, and what they did for me was good.

One land gave me to (another) land. I set out for Byblos (near Beirut), and I returned to Kedem. I spent half a year there. It was Amusinenshi who brought me back: He was the chief of Upper Retenu. He said to me, "You will be well with me, for you will hear the speech of Egypt." He said this for he knew my reputation. He had heard of my intelligence, for the people of Egypt who were there with him bore witness to me. And he said to me, "Why have you come here? Has anything happened at the capitol?" I said to him, "The King of Upper and Lower

Egypt, Sehetepibre, has proceeded to the horizon, and no one knows what may happen because of this." I then spoke equivocally. When I returned from an expedition in the land of the Tjemehu, one announced (that) to me. My mind vacillated. My heart was not in my body, and it brought me to the ways of flight. (But) no one accused me, no one spat in my face. No reproach was heard, and my name was not heard in the mouth of the town crier. I do not know what brought me to this land. It was like the plan of a god.

Praise of Sesostris I

He said to me, "How shall that land fare without him, that efficient god the awe of whom is throughout the foreign lands like Sakhmet in a year of pestilence?" I said to him so that I might answer him, "To be certain, his son has entered the palace and has assumed the inheritance of his father. He is a god, indeed, without peer. No other came into being before him. He is master of knowledge, excellent in planning and efficient in commanding, one by whose command one comes forth and goes down. It was he who controlled the foreign lands while his father was in the palace. He reported to him what he (the father) decreed had come to happen. He is a champion who acts with his own arms, a fighter without anyone like him when he is seen attacking the bowmen and engaging the fray. He is one who bends back the horn and renders hands powerless, so that his enemies cannot muster their ranks. He is vengeful when he cracks skulls, and no one stands up near him. He steps wide when he annihilates the fugitive. There is no chance for the one who shows his back to him. He is upright of heart at the moment of contact. He comes again and does not show his back. He is stalwart of heart when he sees a crowd, and does not allow cowardice around him. He is eager when he falls upon the retinue, and he is joyful when he plunders the bow-people. As soon as he takes up his shield, he strikes down. He need not repeat the act of killing, for there is no one who can deflect his arrow nor one who can draw his bow. The bowmen retreat before him as if before the might of a great goddess. He fights having foreseen the outcome, and he takes no care for the remnants. But he is well-favored and very gentle; through love he takes. His city desires him more than herself, and his people rejoice because of him more that for their gods. Now that he is king, husbands and wives rejoice because of him. While still in the egg, he conquered, and his face was set to it from birth. Those who were born with him have become enriched, for he is one whom the gods have given. How joyful is this land now that he has reigned. He is one to extend borders. He shall vanquish the southlands, and he will not even have to think about the northlands, for he was made to smite the Asiatics and walk over the sandcrossers. Write to him, and let him know your name. Do not cast a spell against his Majesty, for he will not fail to do good to a foreign land which is loyal to him."

Sinuhe in Palestine

And he said to me, "Indeed, Egypt is fortunate, now that she knows that he flourishes. You are here, and you shall be with me, and what I shall do for you will be good." He placed me in front of his children, and he married me to his eldest daughter. He allowed me to pick from his country the choicest part of what he owned on his border with another country.

It was a wonderful land called Yaa. There were cultivated figs in it and grapes, and more wine than water. Its honey was abundant, and its olive trees numerous. On its trees were all varieties of fruit. There were summer corn and barley, and there was no end to all varieties of cattle.

That which fell to my lot as a favored one was great. He set me up as chief of a tribe of the finest in his land. I obtained rations as daily disbursements and wine as a daily requirement, cooked meat and roasted birds, beside the desert game. They hunted for me and they set (food) down before me, in addition to the catch of my hunting dogs. They made for me many sweet things with milk in everything cooked. I spent many years while my offspring became strong men, each man managing his tribe.

The messenger who came north and went south to the capital stayed with me, and I made all Egyptians stay. I gave water to the thirsty man, and I put the wanderer back on the road. I rescued the man who was robbed. When the Asiatics began to stir and to oppose the authority of the chiefs of the foreign lands, I counseled their marches. This ruler of Retenu had me spend many years as an officer of his troops. As to any land which I left, when I had made my attack it was driven off from its cultivation and wells. I had plundered its cattle and brought back its inhabitants, and their produce was taken. I killed the people in it by my strong arm, my bow, my maneuvers, and my efficient advice. It went well with me in his favor, for he loved me and he recognized my bravery. He placed me at the head of his offspring when he saw my arms grow so strong.

The Combat

There came a strong man of Retenu to challenge me at my tent. He was a champion without equal, and he had defeated all of Retenu. He said that he would fight with me, for he thought to ruin me. He planned to take off my cattle, at the urging of his tribe. But that chief talked with me, and I said, "I do not know him. I am not his friend that I could stride about in his camp. Has it ever happened that I opened his door, or have I scaled his walls? He is jealous, for he sees me carrying out your affairs. I am like a bull of a grazing herd in the midst of another herd. The bull of the kine attacks him, but the (Egyptian) bull prevails against him. Is a subject loved when he acts the master? There is no foreign bowman who is an ally of a Delta man. What is it that can join a papyrus plant to a rock? Does a bull wish to show his back through fear of one who might equal him. But if he wishes to fight, let him say so. Is god ignorant of what he has ordained, knowing (as he does) how the matter stands?"

I spent the night stretching my bow and I shot my arrows. I gave play to my dagger, and I fixed up my weapons. When daybreak came, Retenu had come. He had urged on its tribes, and he had collected the lands of both its halves. He had intended this combat. He (the strong man) came out to me where I was waiting, and I placed myself near him. Every heart burned for me, men and women yelled. Every heart was sick for me, saying, "Is there another strong man who could fight him?" He (took up) his shield, his axe, and his armful of javelins. But after I had come away from his weapons, I made his remaining arrows pass by me, as one was not close enough to the other. Then he let out a yell, for he thought to ruin me, and he approached me. I shot him, my arrow fixed in his neck. He shouted and fell upon his nose. I felled him with my axe. I yelled my war cry over his back. Every Asiatic yelped. I gave praise to Montu, and his people celebrated victory. This ruler, Amusininshi, took me in his arms, and kissed me in my clasp. I brought away his possessions, I seized his cattle. What he had thought to do to me I did to him. I took away what was in his tent. I uncovered his camp, and it was abundant for me therein. I became rich in treasure, a great proprietor of cattle.

God acts in such a way to be merciful to one whom he had blamed, one whom he causes to go astray to another land. For today his heart is appeased. A fugitive fled from his haunts, but my renown is in the capital. A wanderer wandered through hunger, but I give bread to my neighbor. Through nakedness a man departed from his land, but I have white clothes and fine linen. A man hurried for lack of something to send, but I gave many servants. My house is fine, and my dwelling place is wide. The thought of me is in the palace.

Sinuhe wishes to return to Egypt

O gods, whoever you are, who decreed this flight, may you be merciful and may you set me in the capital. Perhaps you will let me see the place where my desire lives. What can be more important than joining my dead body to the land where I was born? Come, help me! What has come to pass so far has been good. God has given me satisfaction. May he act similarly to better the end of one whom he had made miserable and be concerned about one whom he had shunted off to live in a foreign land. Today he is merciful, and he hearkens to the prayer of a man far off. May he change my region whence I roamed the earth for him to the place from which he brought me.

The King of Egypt is merciful to me, and I live on his bounty. May I greet the mistress of the land who is in his palace, and may I attend to the errands of her children. My body will be youthful again. Old age has come down on me and feebleness has hurried upon me. My eyes are heavy, and my arms are immobile. My feet stumble, and my senses are exhausted. I am ready for passing on, when they shall send me to the cities of eternity. But may I still serve the Mistress-of-All that she may say something good for me to her children. May she pass eternity above me.

Now this report was made to His Majesty, the King of Upper and Lower Egypt, Kheperkare, who will be judged right, concerning this state in which I was. His Majesty sent to me with provisions of the royal bounty. He rejoiced the heart of this servant as might be done for a ruler of a foreign land. And the king's children who were in his palace had me hear their messages.

The Royal Edict

Copy of the decree brought to this servant regarding his being brought back to Egypt: The Horus Life of Births, the Two Ladies Life of Births, the King of Upper and Lower Egypt, Kheperkare, Son of Re Sesostris, living forever. The king orders Sinuhe: This decree of the king is brought to you to tell you that you have traversed the foreign countries and have come forth from Kedem to Retenu. By your heart's counsel to you, land has given you land. What have you done that one should act against you? You have not blasphemed that one should reprove your words. You have not spoken in the council of the elders that one should restrain your speech. This idea, it took over your senses, although there was nothing in my mind against you. This heaven of yours, which is in the palace, she is well and she flourishes today as in her former state in the kingship of the land, with her children in the audience hall. You shall pile up the treasures they give you, and you shall live off their bounty. When you have come to Egypt, you shall see the capital in which you were born. You shall kiss the ground at the Great Double Gate, and you shall associate with the Companions. Today old age has begun for you, and potency has left you. You have thought about the day of burial, the passing over to an honored state. The night will be appointed for you with oils and poultices from the arms of Tayet (goddess of weaving). A procession will be made for you on the day of internment, the anthropoid sarcophagus (overlaid) with gold leaf, the head with lapis lazuli, and sky above you as you are placed in the outer coffin and dragged by teams of oxen preceded by singers. The dance of the Muu will be performed at your tomb, and the necessary offerings will be invoked for you. They will slaughter at the entrance of your tomb chapel, its pillars to be set up in limestone as is done for the royal children. You shall not die in a foreign land, and Asiatics will not escort you. You shall not be placed in a ram's skin as they make your grave. All of this is too much for one who has roamed the earth. Take thought for your dead body and return.

Sinuhe's Reaction and his Reply

It was while I was standing in the midst of my tribe that this decree reached me; it was after I had prostrated myself and touched the ground that it was read to me. I spread it out over my chest. Then I went about my encampment rejoicing and saying: How could such a thing be done for a servant whose senses led him astray to the land of the barbarians? Indeed, (your) benevolence is excellent, O you who have saved me from Death. Your Ka will allow me to spend the end of my life with my body in the capital.

Copy of the reply to this decree. The servant of the palace Sinuhe says: In peace, in peace. This flight which this servant did in his ignorance is well known by your Ka, O good god, Lord of the Two Lands, whom Re loves and whom Montu, Lord of Thebes, favors, as well as Amun, Lord of the Thrones of the Two Lands, Sobk-Re, Lord of Sumenu, Horus, Hathor, Stum and his Ennead, Sopdu, Neferbau, Semeru, Horus the Easterner, the Mistress of Yemet, may she enfold your head, the council upon the flood waters, Min-Horus in the midst of the desert lands, Wereret, the Mistress of Punt, Nut, Haoreis-Re, and gods who are the Lords of the Beloved Land and the Islands of the Sea. They give life and prosperity to your nostrils; they grant you their bounty. They give you eternity without end and everlastingness without limit. Fear of you is repeated in the lowlands and in the highlands, for you have conquered all that the sun disk encircles. Such is the prayer of this servant to his lord who has rescued him from the West.

Lord of perception, who perceives the people, may he perceive in the Majesty of the palace that this servant was afraid to speak. It is a heavy matter to repeat. The great god, equal of Re, knows the mind of one who has worked for him [of his own accord]. For this servant is in the hands of someone who takes thought for him; I am set in his guidance. Your Majesty is the conquering Horus; your arms prevail over all lands. May now your Majesty command that there be brought Meki from Kedem, Khentiuvash from out of Keshu, and Menus, those who set our authority over the lands of the Fenkhu. They are rulers whose names are worthy and who have been brought up in your love. Not to mention Retenu, for it belongs to you even as your dogs. This which your servant made, it was not premeditated. It was not in my mind. I did not prepare it. I cannot say what separated me from my place. It was like a dream: as when a Delta man sees himself in Aswan or a man of the marshlands in Nubia. Yet I was not afraid. No one chased me. I did not hear a word of censure. No one heard my name in the mouth of the town crier. Except that my body cringed, my feet hastened, and my senses overwhelmed me, with the god who decreed this flight drawing me on. I am not stubborn [in advance]. A man is modest when his homeland is known, for Re has placed the fear of you throughout the land and the dread of you in every foreign land. Whether I am in the capital or in this place, yours is everything that is covered by this horizon. The sun disk rises at your bidding, and the water of the river is drunk if you wish. The air of the heavens is breathed if you speak. Now that this servant has been sent for, this servant will hand over (his property) to his children, whom he has brought up in this place. May your Majesty act as he wishes, for one lives by the air that you give. Re, Horus, and Hathor love your noble nostril; Montu, Lord of Thebes, wishes that they live forever.

Sinuhe's Return

I was allowed to spend a day in Yaa to transfer my goods to my children. My eldest son was in charge of my tribe. My tribe and all my possessions were in his hands, as well as all my serfs, my cattle, my fruit, and all my productive trees. This servant proceeded south. I tarried at the Ways of Horus. The commander in charge of the patrol there sent a message to the capital to give them notice. His Majesty had them send a capable overseer of field laborers of the royal estate and with him ships laden with presents of the royal bounty for the Asiatics who had come with me to lead me to the Ways of Horus. I called each one of them by name. I started out and raised sail. Each servant was at his task. (Dough) was kneaded and strained (for beer) beside me until I reached the wharf of Itjtowy.

Sinuhe at the Palace

When dawn came and it was morning, I was summoned. Ten men came and ten men went to usher me to the palace. I touched my forehead to the ground between the sphinxes. The royal children waited in the gateway to meet me. The Companions who showed me into the pillared court set me on the way to the reception hall. I found His Majesty upon the Great Throne set in a recess (paneled) with fine gold. As I was stretched out on my belly, I lost consciousness in his presence. This god addressed me in a friendly way, and I was like a man caught by nightfall. My soul fled and by body shook. My heart was not in my body: I could not tell life from death.

His Majesty said to one of these Companions: "Lift him up and let him speak to me." And His Majesty said: "See, you have returned, now that you have roamed the foreign lands. Exile has ravaged you; you have grown old. Old age has caught up with you. The burial of your body is no small matter, for now you will not be escorted by the bowmen. Do not creep any more. You did not speak when your name was called out." I feared punishment, and I answered with a timorous answer: "What has my lord said to me? If I try to answer: there is no shortcoming on my part toward god. Fear is in my body, like that which brought to pass the fated flight. I am in your presence. Life belongs to you. May Your Majesty do as he wishes."

The royal children were then brought in, and His Majesty said to the queen: "Here is Sinuhe, who has returned as an Asiatic whom the Bedouin have raised." She let out a cry, and the royal children shouted all together. They said before His Majesty: "It is not really he, O Sovereign, my lord." His Majesty said: "It is he indeed." Then they brought their menyat-necklaces, their rattles, and their sistra with them, and they offered them to His Majesty. "May your arms reach out to something nice, O enduring king, to the ornaments of the Lady of Heaven. May the Golden One give life to your nostrils, and may the Lady of the Stars be joined to you. The crown of Upper Egypt will go northward, and the crown of Lower Egypt will go southward that they may unite and come together at the word of Your Majesty, and the cobra goddess Wadjet will be placed on your forehead. As you have kept your subjects from evil, so may Re, Lord of the Lands, be compassionate toward you. Hail to you. And also to the Lady of All. Turn aside your horn, set down your arrow. Give breath to the breathless. Give us this happy reward, this Bedouin chief Simehyet, the bowman born in Egypt. It was through fear of you that he took flight and through dread of you that he left the land. Yet there is no one whose face turns white at the sight of your face. The eye which looks at you will not be afraid." His Majesty said: "He shall not fear, he shall not be afraid. He shall be a Companion among the nobles and he shall be placed in the midst of the courtiers. Proceed to the audience hall to wait upon him."

Sinuhe Reinstated

When I came from the audience hall, the royal children gave me their hands, and we went to the Great Double Gate. I was assigned to the house of a king's son. Fine things were in it, a cooling room in it, and representations of the horizon. Valuables of the treasury were in it. Vestments of royal linen were in every apartment, and first-grade myrrh of the king and the courtiers whom he loves. Every domestic servant was about his prescribed task. Years were caused to pass from my body. I was depilated, and my hair was combed out. A load was given to the desert, and clothes to the sand-dwellers. I was outfitted with fine linen and rubbed with the finest oil. I passed the night on a bed. I gave the sand to those who live on it and wood oil to those who rub themselves with it. A house of a [plantation owner], which had belonged to a Companion, was given to me. Many craftsmen had built it, and all its trees were planted anew. Meals were brought from the palace three and four times a day, in addition to what the royal children gave. There was not a moment of interruption. A pyramid of stone was built for me in the midst of the pyramids. The overseer of stonecutters of the pyramids marked out its ground plan. The master draftsman sketched in it, and the master sculptors carved in it. The overseers of works who were in the necropolis gave it their attention. Care was taken to supply all the equipment that is placed in a tomb chamber. Ka-servants were assigned to me, and an endowed estate was settled on me with fields attached, at my mooring place, as is done for a Companion of the first order. My statue was overlaid with gold leaf, its apron in electrum. His Majesty ordered it to be done. There was no commoner for whom the like had ever been done. So I remained in the favor of the king until the day of mooring came.

Its beginning has come to its end, as it has been found in writing.
(The traditional colophon marks the end of the story.)

"The Story of Sinuhe." The Literature of Ancient Egypt. Yale University, 1972, pgs 57 - 74.

1. The inhabitants of Palestine and Syria are designated in this text as the Amu, the Setyu, and Pedjtyu (bowmen); the first two are rendered as "Asiatics."
2. Sinuhe identifies himself as an official of Queen Nofru, daughter of Ammenemes I, and wife of his son and successor, Sesostris I. Ka-nofru and Khnumet-sut are respectively the pyramid residence towns of these two first rulers of Dynasty 12.
3. The Tjemehu and Tjehenu, people living to the west of Egypt, are Libyan tribes.
4. Upper Retenu-parts of Palestine and Syria
5. The Ways of Horus was a frontier station on the border of Egypt.
6. Itjtowy was the landing place of the capital, the residence city of the king.

The Tale of the Doomed Prince
Egyptian Tomb Inscription

Once upon a time there was a king, so the story goes, to whom no son had ever been born. But when His Majesty requested a son for himself from the gods of his time, they ordered a birth to be granted him, and he went to bed with his wife in the night. Now when she had became pregnant and had completed the months of childbearing, a son was thus born.

Presently the Hathors came to determine a fate for him and said: He shall die through a crocodile, or a snake, or even a dog. Thus, the people who were at the boy's side heard and then reported it to His Majesty. There upon His Majesty became very much saddened. Then His Majesty had [a house] of stone built [for him] upon the desert, supplied with personnel and with every good thing of the palace, so that the boy did not [need to] venture outside.

Now after the boy had grown older, he went up onto his roof and spied a greyhound following a grownup walking along the road. He said to his servant, who was beside him. What is it that is walking behind the grownup coming along the road? And he told him: It is a greyhound. And the boy told him: Have one like it obtained for me. Thereupon the servant went and reported it to His Majesty. Then His Majesty, said, "Let a young springer be taken to him [because of] his heart's disquiet." And so someone [caused] the greyhound to be taken to him.

Now after days had elapsed upon this, the boy matured in all his body, and he sent to his father saying: What will the outcome be while I am dwelling here? For look, I am committed to Fate. Let me be released so that I may act according to my desire until god does what is his will.

Then a chariot was yoked for him, equipped [with] all sorts of weapons, and [a servant was put in] his following for an escort. He was ferried over to the eastern tract and told: Now you may set out as you wish, [while] his greyhound was with him. He went northward over the desert, following his inclination and living on every sort of desert game.

Presently, he reached the Prince of Nahrin. Now none had been born to the Prince of Nahrin except a marriageable daughter. There had been built for her a house whose window was seventy cubits distant from the ground, and he sent for all the sons of all the princes of the land of Knor and told them: As for the one who will reach the window of my daughter, she shall be a wife for him.

Now after many days had elapsed upon this and while they were [engaged] in their daily practice, presently the boy passed by them. They took the boy to their house, cleansed him, gave fodder to his team, did every sort of thing for the boy, salving him and bandaging his feet, and gave food to his escort. They said to him by way of conversation: Where have you came from, you handsome lad? He told them: I am the son of a chariot warrior of the land of Egypt. My mother died, and my father took for himself another wife, a stepmother. She came to despise me, and [I] left her presence in flight. And they embraced him and kissed him over [all his] body.

[Now after many days had elapsed upon] this, he said to the boys: What is this that you have become engaged in, [boys? And they told him: It has been three] full [months] till now that we have spent time here [leaping up, for the one who] will reach [the] window of the daughter of the Prince of Nahrin, [he will] give her to him for [a wife. He] said to them: If I could but enchant my feet, I would proceed to leap up in your company. They proceeded to leap up according to their daily practice, and the boy stood by afar off observing, while the eyes of the daughter of the Prince of Nahrin were upon him.

Now after [some while] had elapsed upon this, the boy came in order to leap up along with the children of the princes. He leapt up and reached the window of the daughter of the Prince of Nahrin. And she kissed him and embraced him . . . Then someone went in order to impart the news to her father and told him: Somebody has reached the window of your daughter. Then the prince inquired about him saying: The son of which of the princes is he? And he was told: He is a chariot warrior's son. It was from his stepmother's presence that he came in flight from the land of Egypt. Thereupon the Prince of Nahrin became very much angered. He said: Is it to the Egyptian fugitive that I should give my daughter? Send him back home.

And someone came to tell him: "Please set out for the place whence you came." But the daughter seized hold of him and swore by god, saying: "By Pre-Harakhti, if he is taken away from me, I shall neither eat or drink but shall die right away." Then the messenger went and reported to her father every [word] that she had said, and her [father] sent men to slay him while he was still where he was. But the daughter said to [them]: "By Pre, if he is slain, as soon as the sun sets, I shall be dead. I will not stay alive an hour longer than he."

Then [someone went] to tell it to her father. And [her father had] the [lad] and his daughter [brought before] him. The lad [came before] him, and his worth impressed the prince. He embraced him and kissed him over all his body, and he said to him: "Tell me your background. See, you are [now] a son in my eyes." And he told him: "I am the son of a chariot warrior of the land of Egypt. My mother died, and my father took for himself another wife. She came to despise me, and I left her presence in flight." Then he gave him his daughter for a wife and gave him house and fields as well as cattle and all sorts of good things.

Now after [some while] had elapsed upon this, the lad told his wife: I am committed to three fates: crocodile, snake, and dog. Then she told him: Have the dog that follows you killed. And he told her: What a demand! I will not let my dog, which I reared when it was a puppy, be killed. And she came to guard her husband very carefully, not letting him venture outside alone.

Now from the day that the boy had come from the land of Egypt in order to travel about, the crocodile had been his fate . . . It appeared [from the midst of] the lake opposite him in the town in which the lad was [living] with [his wife]. However, a water spirit was in it. Neither would the water spirit let the crocodile emerge nor would the crocodile let the water spirit emerge to stroll about. As soon as the sun rose, [they] both [would be] engaged there in fighting each and every day for a period of three months.

Now after some days had elapsed upon this, the lad sat down and made holiday in his house. And after the end of the evening breeze the lad lay down upon his bed, and slumber took possession of his body. Then his wife filled one jar [with wine and filled] another jar with beer. Presently a [snake] emerged [from its] hole to bite the lad, but his wife was sitting beside him without going to sleep. The [jars were thus standing] accessible to the snake, and it imbibed and became intoxicated. Then it reclined and turned upside down. Thereupon [his wife caused] it to be [split] into segments with her hand-axe. She then awoke her husband. . ., and she told him: See, your god has delivered one of your fates into your hand. He will guard [you henceforth. Then he] made an offering to Pre, praising him and extolling his power daily.

Now after [some days had elapsed upon this], the lad went out to stroll about for relaxation on his property. [His wife] did not go out [with him], but his dog was following him. Then his dog took a bite, [saying: I am your fate. Thereupon] he fled before it. Presently he reached the lake and descended into the [water in flight before the] dog. And so the crocodile [seized him] and carried him off to where the water spirit [usually] was, [but he was not there.

[The] crocodile told the lad: I am your fate who has been fashioned so as to come in pursuit of you, but [it is three full months] now that I have been fighting with the water spirit. See, I shall let you go. If my [opponent returns to engage me] to fight, [come] and lend me your support [in order to] kill the water spirit. But if you . . see the crocodile.

Now after dawn and the next day had come about, the [water spirit] returned. . .
(The remainder of the tale is lost.)

"The Tale of the Doomed Prince." <u>The Literature of Ancient Egypt</u>. Yale University, 1972
According to popular religious belief in the New Kingdom, there were seven such Hathor goddesses, who determined the fate of a child at birth.

Theseus and the Minotaur

Theseus was the son of the Athenian king, Aegeus. He spent his youth in his mother's home, a city in southern Greece. Aegeus went back to Athens before the child was born, but first he placed in a hollow a sword and a pair of shoes and covered them with a great stone. He did this with the knowledge of his wife and told her that whenever the boy - if it was a boy - grew strong enough to roll away the stone and get the things beneath it, she should send him to Athens to claim him as his father. The child was a boy and he grew up strong far beyond others, so that when his mother finally took him to the stone he lifted it with no trouble at all. She told him then that the time had come for him to seek his father, and a ship was placed at his disposal by his grandfather. But Theseus refused to go by water because the voyage was safe and easy. His idea was to become a great hero as quickly as possible.

After many adventures Theseus arrived in Athens and was invited to a banquet by the king, who was not aware that Theseus was his son. The king was actually fearful of Theseus' popularity, thinking the people might make him their king. Theseus wished to make himself known to his father. He drew out the sword that Aegeus instantly recognized. Aegeus then proclaimed to the country that Theseus was his son and heir.

Years before his arrival in Athens, a terrible misfortune had happened to the city. Minos, the powerful ruler of Crete, had lost his only son, Androgeus, while the young man was visiting the Athenian king. King Aegeus had done what no host should do; he had sent his guest on an expedition full of peril -- to kill a dangerous bull. Instead, the bull had killed the youth. Minos invaded the country, captured Athens and declared that he would raze it to the ground unless every nine years the people sent him a tribute of seven maidens and seven youths. A horrible fate awaited these young people. When they reached Crete they were given to the Minotaur to devour.

The Minotaur was a monster, half bull, half human, the offspring of Minos' wife Pasiphae and a wonderfully beautiful bull. Poseidon had given this bull to Minos in order that he should sacrifice it to him, but Minos could not bear to slay it and had kept it for himself. To punish him, Poseidon had made Pasiphae fall madly in love with it.

When the Minotaur was born Minos did not kill him. He had Daedalus, a great architect and inventor, construct a place of confinement for him from which escape was impossible. Daedalus built the Labyrinth, famous throughout the world. Once inside, one would go endlessly along its twisting paths without ever finding the exit. To this place the young Athenians were each time taken and left to the Minotaur. There was no possible way to escape. In whatever direction they ran they might be running straight to the monster; if they stood still he might at any moment emerge from the maze. Such was the doom that awaited fourteen youths and maidens a few days after Theseus reached Athens. The time had come for the next installment of tribute.

At once Theseus came forward and offered to be one of the victims. All loved him for his goodness and admired him for his nobility, but they had no idea that he intended to try and kill the Minotaur. He told his father, however, and promised him that if he succeeded, he would have the black sail which the ship with its cargo of misery always carried changed to a white one, so that Aegeus could know long before it came to land that his son was safe.

When the young victims arrived in Crete they were paraded before the inhabitants on their way to the Labyrinth. Minos' daughter Ariadne was among the spectators and she fell in love with Theseus at first sight as he marched past her. She sent for Daedalus and told him he must show her a way to get out of the Labyrinth, and she sent for Theseus and told him she would bring about his escape if he would promise to take her back to Athens and marry her. He agreed and she gave him the clue she had received from Daedalus, a ball of thread that he was to fasten at one end to the inside of the door and unwind as he went on. This he did, and certain that he could retrace his steps whenever he chose, he walked boldly into the maze looking for the Minotaur. He came upon him asleep and fell upon him, pinning him to the ground and battered the monster to death with his fists.

When Theseus lifted himself up from that terrific struggle, the ball of thread lay where he had dropped it. With it in his hands, the way out was clear. The others followed and taking Ariadne with them they fled to the ship and over the sea to Athens.

In all the excitement of escaping the Minotaur and returning home Theseus forgot to change the sails. His father saw the black sail from the Acropolis, where for days he had watched the sea. It was to him the sign of his son's death and he threw himself into the sea and was killed. This sea was called the Aegean ever after.

So Theseus became king of Athens, a wise but disinterested king. He declared to the people that he did not wish to rule over them; he wanted a people's government where all would be equal. He resigned his royal power and organized a commonwealth, building a council hall where the citizens should gather and vote. The only office he kept for himself was that of Commander in Chief. Thus Athens became, of all earth's cities, the happiest and most prosperous, the only true home of liberty, the one place in the world where the people governed themselves.

Taken from the writings of Apollodorus. first or second century A.D.

Daedalus and Icarus

Daedalus learned his craftsmanship from the gods themselves. He was the architect who contrived the Labyrinth for the Minotaur in Crete. When King Minos learned that the Athenians had escaped and took his daughter, he was convinced that Daedalus had helped them. Accordingly, he imprisoned him and his son Icarus in the Labyrinth.

Daedalus knew that land and sea would be guarded, but what of the air. He watched the birds of the sky and thought of a plan to escape. He collected feathers and candle wax. He made two pairs of wings. He fitted them on and by waving his arms he could fly. He warned Icarus to keep a middle course over the sea; to fly low into the fogs would weigh down the feathers. If he flew too high the heat of the sun might melt the wax and the wings fall apart. As the two flew lightly and without effort away from Crete the delight of this new and wonderful power went to the boy's head. He soared exultingly up and up, paying no heed to his father's anguished commands. Then he fell. The wax softened and came apart. He dropped into the sea and the waters closed over him. All that could be seen were feathers floating on the surface of the sea. The afflicted father flew safely to Sicily, where the king received him kindly.

Minos was enraged at his escape and determined to find him. He made a cunning plan. He had it proclaimed everywhere that a great reward would be given to whoever could pass a thread through an intricately spiraled shell. Daedalus told the Sicilian king that he could do it. He bored a small hole in the closed end of the shell, fastened a thread to an ant, introduced the ant into the hole, and then closed it. When the ant finally came out at the other end, the thread, of course, was running clear through all the twists and turns. "Only Daedalus would think of that," Minos said, and he came to Sicily to seize him. But the Sicilian king refused to surrender him, and in the contest Minos was slain.

Taken from the writings of Apollodorus, first or second century A.D.

The Golden Reed Pipe
(A Yao Folktale)

Once upon a time there lived in the mountains a woman and her daughter. The daughter liked to dress in red. Hence her name, Little Red.

One day they were plowing and sowing in the fields. All of a sudden, a gale blew up and in the sky there appeared an evil dragon who stretched down his claws, caught Little Red in a tight grip and flew off with her towards the west. Her mother vaguely heard her daughter's words carried on the wind:

"Oh mother, oh mother, as dear as can be!

My brother, my brother will rescue me!"

Wiping away her tears, her mother gazed into the sky and said, "But I only have a daughter. Who can this brother be?"

She staggered home and was halfway there when her white hair was caught up in the branches of a bayberry tree growing by the roadside. While she was disentangling her hair, she spotted a red, red berry dangling from a twig. She picked and swallowed it without thinking.

When she arrived home, the woman gave birth to a boy with a round head and red cheeks. She named the boy Little Bayberry.

Bayberry grew up very quickly and in a few days he was a young lad of fourteen or fifteen.

His mother wanted to ask Bayberry to rescue his sister but couldn't bring herself to inflict such a dangerous task on him. All she could do was weep to herself in secret.

One day a crow alighted on the eaves of her house and cried:

"Your sister's suffering out there, out there!

She's weeping in the Evil Dragon's lair!

Bloodstains on her back,

She's digging rocks with hands so bare!"

Upon hearing this, Bayberry asked his mother:

"Do I have a sister?"

Tears streaming down her cheeks, his mother replied, "Yes, my boy, you do. Because she loved to dress in red, she was called Little Red. That Evil Dragon who has killed so many people came and took her away."

Bayberry picked up a big stick and said, "I'm going to rescue Little Red and kill that Evil Dragon. Then he can't do any more harm!"

His mother leaned against the doorframe and through misty eyes watched her son march away.

Bayberry walked for miles and miles. On a mountain road he saw ahead of him, blocking the way, a large rock. It was pointed and rubbed smooth by all the travelers who had had to climb it. One wrong step would mean a nasty fall.

Bayberry said, "This is my first obstacle! If I don't remove it now, it will be the undoing of many more people." He thrust his stick under the rock and heaved with all his might. There was a great "Crack!" and the stick broke in two. Then he put both his hands under the rock and tried to shift it with all his strength. The rock rolled down into the valley.

Just at that moment, a shining golden reed pipe appeared in the pit where the rock had been. Bayberry picked it up and blew on it. It gave out a resonant sound.

Suddenly, all the earthworms, frogs, and lizards by the roadside began to dance. The quicker the tune, the faster the creatures danced. As soon as the music stopped, they ceased dancing. Bayberry had an idea: "Ah! Now I can deal with the Evil Dragon."

He strode away, the golden reed pipe in hand. He climbed a huge rocky mountain and saw a ferocious-looking dragon coiled at the entrance to a cave. Piles of human bones lay all around him. He also saw a girl in red chiseling away at the cave. Tears were streaming down her cheeks. The Evil Dragon whipped the girl on the back with his tail and shouted vilely at her:

"Most ungrateful loathsome Mistress Red!
Since with me you would not wed,
Day by day,
Rock by rock,
Hew me out a handsome cave,
Or I'll send you to your grave!"

Bayberry realized that the girl was none other than his sister. He shouted:

"Wicked monster! Evil fiend!
To torment my sister so!
Till your wretched life shall end
On this pipe I'll blow!"

Bayberry began to blow on his golden reed pipe. The music set the Evil Dragon dancing despite himself. Little Red downed her chisel and emerged from the cave to watch.

Bayberry blew on the pipe. The Evil Dragon continued to dance, squirming and writhing. The quicker the tune, the faster the Evil Dragon moved.

Little Red came over and wanted to speak to her brother. With a gesture of his hand, Bayberry showed her that he could not stop playing the pipe. If he did, the Evil Dragon would eat them both up.

Bayberry kept blowing for all he was worth, and the Evil Dragon stretched his long waist and kept writhing around in time to the music.

Fire came from his eyes, steam from his nostrils, and panting breath from his mouth. The Evil Dragon pleaded:

"Ho-ho-ho! Brother you're the stronger!
Blow no more! Torture me no longer!
I'll send her home,
If you leave me alone!"

Bayberry had no intention of stopping. As he blew, he walked towards a big pond. The Evil Dragon followed him to the bank of the pond, squirming and dancing all the way. With a great splash the Evil Dragon fell into the pond and the water rose several feet. The Evil Dragon was utterly exhausted. Fire came from his eyes, steam from his nostrils and panting breath from his mouth. He entreated again in a hoarse voice:

"Ho-ho-ho! Brother you're the stronger!
Let me alone and I'll stay in this pond
And torture folk no longer!"

Bayberry replied:

"Wicked fiend!
This is my bargain:
Stay at the bottom of this pond,
And never do harm again."

The Evil Dragon kept nodding his head.   As soon as the golden reed pipe stopped blowing, he sank to the bottom of the pond.

Bayberry took hold of his sister's hand and walked happily away.

Not long after they set off, they heard the sound of water splashing in the pond. They looked over their shoulders and saw the Evil Dragon emerge from the water pond. He raised his head and flew in their direction, baring his fangs and clawing the air.

Little Red cried:
>"Go deep when digging a well;
>Pull up the roots when hoeing a field.
>While that Dragon is still alive
>To kindly ways he'll never yield."

Bayberry rushed back to the pond and began to blow on his pipe once more. The Evil Dragon fell back into the pond and began to dance again, squirming and writhing in the water.

Bayberry stood on the bank for seven days and nights, a fast tune blowing on his pipe. Finally, the Evil Dragon could move no longer and floated on the surface of the water. His days had come to an end.

Sister and brother joyfully returned home, dragging the body of the Evil Dragon along behind them. When their mother saw her two children coming home, her face lit up with happiness.

They peeled the dragon's skin to make a house, took out the dragon's bones to serve as pillars and beams and cut off the dragon's horn to make ploughshares. With the dragon's horn they ploughed the fields quickly and had no need of oxen. In this way they ploughed many fields, sowed much grain and enjoyed a life of plenty.

Yao Folktale   pgs 161-167

The Vulnerable Spot
(A Yi Folktale)

Once upon a time there was a man digging the ground on the slope of a mountain, and a witch wanted to eat him up. She sat in front of him and riveted her eyes on him. When the man saw the witch staring at him so, he said to her, "Don't eat me today. At home I have a speckled hen. You can have me after I have eaten the hen and become fat."

The witch agreed.

So the man killed his speckled hen and ate it that evening, keeping a piece of meat which he took to the witch the next day.

The witch ate the meat and said, "Now I must eat you."

The man said to her, "You'd better not eat me today. At home I have a sow that is so fat that its dugs trail on the ground. You can have me when I have eaten her and become even fatter!"

The witch agreed again.

So that evening the man ate the fat sow whose dugs trailed on the ground. He kept a chunk of pork and gave it to the witch the next day.

The witch ate the pork and said, "I'm full today. I shall eat you tomorrow."

The man went home, looked all around, but could find nothing more to eat. In despair he went to his neighbors and said, "The witch wants to eat me up. On the first day, I told her that she could eat me after I had eaten the speckled hen and became a little fatter. So she spared my life. On the second day, I told her not to eat me until I had eaten my fat sow whose dugs trailed on the ground. So she spared my life again. But now, since I have neither my speckled hen nor my fat sow, the witch is bound to eat me tomorrow. Good neighbors, please take care of my wife and children."

His neighbors consoled him, "Don't worry! Don't worry! Let's all work out a way to kill the witch tomorrow."

So they made a plan together and promised to meet on the mountain slope the following day.

The next morning the man came to dig the ground on the mountain slope as usual. The witch was already waiting for him. As she was afraid to be seen, she had transformed herself into a tree stump and sat in front of the man. When the neighbors arrived and saw the stump, they asked the man, "Tell us what is this in front of you?"

The witch heard this and she said to the man,
"Tell them it is a tree stump."

The man copied the witch's tone of voice.

"Oh! Tell them it is a tree stump."

The witch immediately corrected him, "No! Say it in your own voice."

The man repeated, "No! Say it in your own voice."

The neighbors heard this and said, "What do you mean by these words? We don't believe you. If it is a stump then chop it with your hoe to prove it,"

The witch cried, "Chop me! Hold your hoe up high but swing it down lightly. Don't chop my vulnerable spot."

The man said, "Oh, Grannie, where is your vulnerable spot? Show me, or else I might chop it by mistake."

"That's it there! said the witch trembling with fear. "That black mark. That's my vulnerable spot. Whatever you do, don't chop it by mistake."

"Of course not," replied the man, "of course I won't." He raised his hoe high above his head and brought it down with all his might on the witch's vulnerable spot. The witch died instantly.

The Legend of Silk

No one knows for sure when silk was discovered. According to a Chinese legend, it was discovered about 2700 B.C. in the garden of Emperor Huang-Ti. The Emperor ordered his wife, Hsi-Ling-Shi, to find out what was damaging his mulberry trees.

Hsi-Ling-Shi found white worms eating the mulberry leaves and spinning shiny cocoons. She accidentally dropped a cocoon into hot water. As she played with the cocoon in the water, a delicate, cobwebby tangle separated itself from the cocoon. Hsi-Ling-Shi drew it out and found that one slender thread was unwinding itself from the cocoon. She had discovered silk.

Hsi-Ling-Shi persuaded her husband to give her a grove of mulberry trees where she could grow thousands of worms that spun such beautiful cocoons. It is said that Hsi-Ling-Shi invented the silk reel, which joined these fine filaments into thread thick and strong enough for weaving.

No one knows how much, if any, of the story is true. But historians do know that silk was first used in China. The Chinese guarded the secret of the silk worm. Disgrace and death faced the traitor who disclosed the origin of silk to the outside world. Only the Chinese knew how to make silk for about 3,000 years.

from Rig Veda
Aryan Hymns

10.81   The All-Maker

1. The sage, our father, who took his place as priest of the oblation and offered all these words as oblation, seeking riches through prayer, he entered those who were to come later, concealing those who went before.
2. What was the base, what sort of raw matter was there, and precisely how was it done, when the All-Maker, casting his eye on all, created the earth and revealed the sky in its glory?
3. With eyes on all sides and mouths on all sides, with arms on all sides and feet on all sides, the One God created the sky and the earth, fanning them with his arms.
4. What was the wood and what was the tree from which they carved the sky and the earth? You deep thinkers, ask yourselves in your own hearts, what base did he stand on when he set up the worlds?
5. Those forms of yours that are highest, those that are lowest, and those that are in the middle, Oh, All-Maker, help your friends to recognize them in the oblation. You who follow your own laws, sacrifice your body yourself, making it grow great.
6. All-Maker, grown great through the oblation, sacrifice the earth and sky yourself. Let other men go astray all around; let us here have a rich and generous patron.
7. The All-Maker, the lord of sacred speech, swift as thought - we will call to him today to help us in the contest. Let him who is the maker of good things and is gentle to everyone rejoice in all our invocations and help us.

10. 135 The Boy and the Chariot

1. (The son:) 'Beneath the tree with beautiful leaves where Yama drinks with the gods, there our father, the head of the family, turns with longing to the ancient ones.
2. 'Reluctantly I looked upon him as he turned with longing to the ancient ones, as he moved on that evil way I longed to have him back again.'
3. (Voice of the father:) 'In your mind, my son, you made a new chariot without wheels, which had only one shaft but can travel in all directions. And unseeing, you climbed into it.
4. 'My son, when you made the chariot roll forth from the priests, there rolled after it a chant that was placed from there upon a ship.'
5. Who gave birth to the boy? Who made the chariot roll out? Who could tell us today how the gift for the journey was made?
6. How was the gift for the journey made? The beginning arose from it: first they made the bottom, and then they made the way out.
7. This is the dwelling-place of Yama, that is called the home of the gods. This is his reed pipe that is blown, and he is the one adorned with songs.

10.117 In Praise of Generosity

1. The gods surely did not ordain hunger alone for slaughter; various deaths reach the man who is well fed. The riches of the man who gives fully do not run out, but the miser finds no one with sympathy.

2. The man with food who hardens his heart against the poor man who comes to him suffering and searching for nourishment - though in the past he had made use of him - he surely finds no one with sympathy.

3   The man who is truly generous gives to the beggar who approaches him thin and in search of food. He puts himself at the service of the man who calls to him from the road, and makes him a friend for times to come.

4. That man is no friend who does not give of his own nourishment to his friend, the companion at his side. Let the friend turn away from him; this is not his dwelling place. Let him find another man who gives freely, even if he be a stranger.

5. Let the stronger man give to the man whose need is greater; let him gaze upon the lengthening path. For riches roll like the wheels of a chariot, turning from one to another.

6. The man without foresight gets food in vain; I speak the truth: it will be his death. He cultivates neither a patron nor a friend. The man who eats alone brings troubles upon himself alone.

7. The plough that works the soil makes a man well fed; the legs that walk put the road behind them. The priest who speaks is better that the one who does not speak. The friend who gives freely surpasses the one who does not.

8. One-foot surpasses Two-foot; and Two-foot leaves Three-foot behind. Four-foot comes at the call of Two-foot, watching over his herds and serving him.

9. The two hands, though the same, do not do the same thing. Two cows from the same mother do not give the same amount of milk. The powers of two twins are not the same. Two kinsmen do not give the same generosity.

A Prayer to the Gods of Night
Old Babylonian Poem

They are lying down, the Great Ones,
The bars have fallen, the bolts are shot,
The crowds and all the people rest,
The open gates are locked.
The gods of the land, the goddesses,
Shamash Sin Adad Ishtar,
Sun, moon, turmoil, love
Lie down to sleep in heaven.
The judgment seat is empty now,
For no god now is still at work.
Night has drawn down the curtain,
The temples and the sanctuaries are silent, dark,
Now the traveler calls to his god,
Defendant and plaintiff sleep in peace,
For the judge of truth, the father of the fatherless,
Shamash, has gone to his chamber.
"O Great Ones, Princes of the Night,
Bright Ones, Gibil the furnace, Irra,
Warlord of the Underworld,
Bow-star and Yoke, Orion, Pleiades, Dragon,
The Wild Bull, the Goat, and the Great Bear,
Stand by me in my divination.
By this lamb that I am offering,
May truth appear!"

From The Code of Hammurabi

--3,4 False witness

If a man has come forward in a lawsuit for the witnessing of false things, and has not proved the thing that he said, if that lawsuit is a capital case, that man shall be put to death. If he came forward for witnessing about corn or silver, he shall bear the penalty (which would apply to) that case.

--16 Theft by finding

If a man has concealed in his house a lost slave or slave-girl belonging to the Palace or to a subject, and has not brought him (or her) out at the proclamation of the Crier, the owner of the house shall be put to death. {It must be understood that during this time there were no civil rights.} The one concealing the slave was doing so not as a matter of conscience but for personal gain, hoping to acquire someone else's slave for nothing.

--25 Looting

If a fire has broken out in a man's house, and a man who has gone to extinguish it has cast his eye on the property of the owner of the house and has taken the property of the owner of the house, that man shall be thrown into the fire.

--48 Moratorium in case of hardship

If a man is subject to a debt bearing interest, and Adad (the Weather-god) has saturated his field or a high flood has carried (its crop) away, or because of lack of water he has not produced corn in that field, in that year he shall not return any corn to (his) creditor. He shall cancel (literally 'wet') his tablet, and he shall not pay interest for that year.

--128 The status 'wife' depends upon a contract

If a man has taken a wife, but has not set down a contract for her, that woman is not (legally) a wife.
Example contract: Puzur-khaya has taken Ubartum as his wife. The oath by the King has been taken before (four named persons), acting as witnesses. Year in which Enamgalanna was installed as En-priest of Inanna.

--165 Inheritance, and the right to bequeath property by will

If a man has donated field, orchard or house to his favorite heir and has written a sealed document for him, after the father has gone to his doom, when the brothers share, he (the favorite heir), shall take the gift that his father gave him, and apart from that they shall share equally in the property of the paternal estate.

--188.189 Adoption and apprenticeship

If an artisan has taken a child for bringing up, and has taught him his manual skill, (the child) shall not be (re)claimed. If he has not taught him his manual skill, that pupil may return to his father's house.

--195-99, 205, 206 Assault

If a son has struck his father, they shall cut off his hand.
If a man has destroyed the eye of a man of the 'gentleman' class, they shall destroy his eye. If he has broken a gentleman's bone, they shall break his bone. If he has destroyed the eye of a commoner or broken a bone of a commoner, he shall pay one mina of silver. If he has destroyed the eye of a gentleman's slave, or broken a bone of a gentleman's slave, he shall pay half (the slave's) price.
If a gentleman's slave strikes the cheek of a man of the 'gentleman' class, they shall cut off (the slave's) ear.
If a gentleman strikes a gentleman in a free fight and inflicts an injury on him, that man shall swear "I did not strike him deliberately", and he shall pay the surgeon. (the man's liability is limited to paying the surgeon.)

--215-18 Fees and penalties connected with surgery

If a surgeon has made a serious wound in a gentleman with a bronze knife, and has thereby saved the gentleman's life, . . . he shall receive ten shekels of silver. If a commoner, he shall receive five shekels of silver. If a gentleman's slave, the slave's master shall pay the surgeon two shekels of silver.
If the surgeon has made a serious wound in a gentleman with a bronze knife, and has thereby caused the gentleman to die, . . they shall cut off (the surgeon's) hand.

--229-30 Treatment of jerry-builders

If a builder has made a house for a man but has not made his work strong, so that the house he made falls down and causes the death of the owner of the house, that builder shall be put to death. If it causes the death of the son of the owner of the house, they shall kill the son of that builder.

The Iliad
List of Characters

THE ACHAIANS

Achilles	The central character of the Iliad and the greatest warrior in the Achaian army. The most significant flaw in his temperament is his excessive pride.
Agamemnon	The well meaning but vacillating king of Mycenae; commander-in-chief of the forces against Troy; he is the brother of Menelaos.
Aias (Ajax)	Son of Telamon, he is often called Telarnonian Aias; his reputation is due primarily to brute strength and courage.
Aias the Lesser	He is a distinguished warrior, but brazen and conceited
Antilochos	The son of Nestor; a brave young warrior who takes an active part in the fighting.
Automedon	The squire and charioteer of Achilles.
Diomedes	One of the finest and bravest of the Achaian warriors - he is always wise and reasonable and is known for his courtesy and gallantry
Helen	She was originally married to Menelaos, but she ran away to Troy with Paris and became his wife. She is supposedly the most beautiful woman in the world; however, she is also self-centered, insincere, and insensitive.
Idomeneus	He is one of the most efficient of the Achaian leaders and has the respect and liking of the whole Achaian army.
Kalchas	Soothsayer and prophet of the Achaians
Menelaos	Husband of Helen; brother of Agamemnon
The Myrmidons	Soldiers of Achilles
Nestor	The oldest of the Achaian warriors having all the wisdom and experience of age. He is a valuable asset in the council. Although he can no longer fight, he remains at the front line of every battle, commanding his troops.
Odysseus	The shrewdest and most subtle of all the Achaians; King of Ithaca
Patroklos	Achillles' close friend and warrior-companion

THE TROJANS AND THEIR ALLIES

Andromache   The wife of Hektor; loyal, loving, and submissive.

Aenas   Son of Aphrodite; a Trojan nobleman. He is second in command of the Trojan army and a brave, skillful warrior.

Antenor   A Trojan nobleman who advocates the return of Helen to the Achaians.

Astyanax   The infant son of Hektor and Andromache.

Chryseis   Daughter of Chryses, the priest of Apollo. She becomes Agarnemnons' "war prize"

Dolon   A Trojan nobleman captured by Odysseus and Diomedes.

Glaukos   A prince and renowned warrior.

Hekuba   Wife of Priam and mother of Hektor.

Hektor   Prince of Troy and son of Priam and Hekuba. He is the greatest of the Trojan warriors and one of the most noble persons in the Iliad.

Helenos   Son of Priam and Hekuba; a prince of Troy and a seer.

Cassandra   Daughter of Priam and Hekuba

Pandaros   A good archer, but a treacherous man; it is he who breaks the truce.

Paris   Son of Priam and Hekuba; a prince of Troy. It was Paris who abducted Helen and brought her to Troy. His reputation is that of a "pretty boy".

Poulydamas   One of the Trojan leaders; a very able and clear-headed military strategist whose advice to Hektor is usually not heeded.

Priam   King of Troy. He is very old and no longer able to command his army in the field, but his great courage is shown in the last book of the Iliad.

THE GODS

Aphrodite Daughter of Zeus; goddess of love. She is the mother of the mortal Aeneas and is the patron of Paris, therefore she fights on the Trojan side.

Apollo Son of Zeus; god of prophecy, light, poetry, and music. He fights on the Trojan side.

Ares Son of Zeus and Hera; god of war. He is the lover of Aphrodite and fights on the Trojan side despite an earlier promise to Hera and Athena that he would support the Achaians.

Artemis Daughter of Zeus; sister of Apollo; goddess of chastity, hunting, and wild animals. She fights on the Trojan side.

Athena Daughter of Zeus; goddess of wisdom. She plays a major role in the war, fighting for the Achaians

Dione Mother of Aphrodite.

Hades God of the dead and ruler of the underworld.

Hera Sister and wife of Zeus. She is the most fanatical supporter of the Achaians and will stop at nothing, including deceiving her husband, to achieve the defeat of Troy.

Hermes Ambassador of the gods, he is on the Achaians' side.

Iris A messenger of the gods.

Poseidon Younger brother of Zeus; god of the sea. He is a strong supporter of the Achaians.

Thetis Mother of Achilles, a sea nymph. She is a staunch advocate of her son and does all she can to help him.

Xanthos Son of Zeus; god of one of the major rivers of Troy. He fights against Achilles in Book 21.

Zeus The supreme god and king of Olympus. His duty is to carry out the will of Destiny, so he is officially neutral in the war, but he is sympathetic toward the Trojans. He supports Achilles against Agamemnon.

The Iliad by Homer
Book One - Quarrel, Oath, and Promise

Anger be now your song, immortal one,
Achilles' anger, doomed and ruinous,
that caused the Achaians loss on bitter loss
and crowded brave souls into the undergloom,
leaving so many dead men -- carrion
for dogs, and birds; and the will of Zeus was done.
Begin it when the two men first contending
broke with one another -

The Lord Marshal
Agamemnon, Atreus' son, and Prince Achilles,
Among the gods, who brought this quarrel on?
The son of Zeus by Leto. Agamemnon
angered him, so he made a burning wind
of plague rise in the army: rank and file
sickened and died for the ill their chief had done
in despising a man of prayer.
This priest, Khryses, had come down to the ships
with gifts, no end of ransom for his daughter;
on a golden staff he carried the god's white bands
and sued for grace from the men of all Achaia,
The two Atreidai most of all:

"O captains
Menelaos and Agamemnon, and you other
Achaians under arms!
The gods who hold Olympus, may they grant you
plunder of Priam's town and a fair wind home,
but let me have my daughter back for ransom
as you revere Apollo, son of Zeus!"

Then all the soldiers murmured their assent:

"Behave well to the priest. And take the ransom!"

But Agamemnon would not. It went against his desire,
and brutally he ordered the man away:

"Let me not find you here by the long ships
loitering this time or returning later,
old man; if I do,
the staff and ribbons of the god will fail you.

Give up the girl? I swear she will grow old
at home in Argos, far from her own country,
working my loom and visiting my bed.
Leave me in peace and go, while you can, in safety."

So harsh he was, the old man feared and obeyed him,
in silence trailing away
by the shore of the tumbling clamorous whispering sea,
and he prayed and prayed again, as he withdrew,
to the god whom silken-braided Leto bore:

"O hear me, master of the silver bow,
protector of Tenedos and the holy towns,
Apollo, Sminthian, if to your liking
ever in any grove I roofed a shrine
or burnt thighbones in fat upon your altar --
bullock or goat flesh -- let my wish come true:
your arrows on the Danaans for my tears!"

Now when he heard this prayer, Phoibos Apollo
walked with storm in his heart from Olympus' crest,
quiver and bow at his back, and the bundled arrows
clanged on the sky behind as he rocked in his anger,
descending like night itself. Apart from the ships
he halted and let fly, and the bowstring slammed
as the silver bow sprang, rolling in thunder away.
Pack animals were his target first, and dogs,
but soldiers, too, soon felt transfixing pain
from his hard shots, and pyres burned night and day.
Nine days the arrows of the god came down
broadside upon the army. On the tenth,
Achilles called all ranks to assembly. Hera,
whose arms are white as ivory, moved him to it,
as she took pity on Danaans dying.
All being mustered, all in place and quiet,
Achilles, fast in battle as a lion,
rose and said:

                "Agamemnon, now, I take it,
the siege is broken, we are going to sail,
and even so may not leave death behind:
if war spares anyone, disease will take him . . .
We might, though, ask some priest or some diviner,
even some fellow good at dreams -- for dreams
come down from Zeus as well --
why all this anger of the god Apollo?

Has he some quarrel with us for a failure
in vows or hecatombs? Would mutton burned
or smoking goat flesh make him lift the plague?"

Putting the question, down he sat. And Kalchas,
Kalchas Thestoides, came forward, wisest
by far of all who scanned the flight of birds.
He knew what was, what had been, what would be,
Kalchas, who brought Achaian's ships to Ilion
by the diviner's gift Apollo gave him.
Now for their benefit he said:

          "Achilles,
dear to Zeus, it is on me you call
to tell you why the Archer God is angry.
Well, I can tell you. Are you listening? Swear
by heaven that you will back me and defend me,
because I fear my answer will enrage
a man with power in Argos, one whose word
Achaian troops obey.

          A great man in his rage is formidable
for underlings: though he may keep it down,
he cherishes the burning in his belly
until a reckoning day. Think well
if you will save me."

Said Achilles:
          "Courage.
Tell what you know, what you have light to know.
I swear by Apollo, the lord god to whom
you pray when you uncover truth,
never while I draw breath, while I have eyes to see,
shall any man upon this beachhead dare
lay hand on you -- not one of all the army,
not Agamemnon, if it is he you mean,
though he is first in rank of all Achaians."

The diviner then took heart and said:
          "No failure
in hecatombs or vows is held against us.
It is the man of prayer whom Agamemnon
treated with contempt: he kept his daughter,
spurned his gifts: for that man's sake the Archer
visited grief upon us and will again.
Relieve the Danaans of this plague he will not
until the girl who turns the eyes of men

shall be restored to her own father -- freely,
with no demand for ransom -- and until
we offer up a hecatomb at Khryse.
Then only can we calm him and persuade him."

He finished and sat down. The son of Atreus,
ruler of the great plain, Agamemnon,
rose, furious. Round his heart resentment
welled, and his eyes shone out like licking fire.
Then, with a long and boding look at Kalchas,
he growled at him:

"You visionary of hell,
never have I had fair play in your forecasts.
Calamity is all you care about, or see,
no happy portents; and you bring to pass
nothing agreeable. Here you stand again
before the army, giving it out as oracle
the Archer made them suffer because of me,
because I would not take the gifts
and let the girl Khryseis go; I'd have her
mine, at home. Yes, if you like, I rate her
higher than Clytaimnestra, my own wife!
She loses nothing by comparison
in beauty or womanhood, in mind or skill.

For all of that, I am willing now to yield her
if it is best; I want the army saved
and not destroyed. You must prepare, however,
a prize of honor for me, and at once,
that I may not be left without my portion --
I, of all Argives. It is not fitting so.
While every man of you looks on, my girl
goes elsewhere."

Prince Achilles answered him:

"Lord Marshal, most insatiate of men,
how can the army make you a new gift?
Where is our store of booty? Can you see it?
Everything plundered from the towns has been
distributed; should troops turn all that in?
Just let the girl go, in the god's name, now;
we'll make it up to you, twice over, three
times over, on that day Zeus gives us leave
to plunder Troy behind her rings of stone."

Agamemnon answered:
                    "Not that way
will I be gulled, brave as you are, Achilles.
Take me in, would you? Try to get around me?
What do you really ask? That you may keep
your own winnings, I am to give up mine
and sit here wanting her? Oh, no:
the army will award a prize to me
and make sure that it measures up, or if
they do not, I will take a girl myself,
your own, or Aias', or Odysseus' prize!
Take her, yes, to keep. The man I visit
may choke with rage; well, let him.
But this, I say, we can decide on later.

Look to it now, we launch on the great sea
a well-found ship, and get her manned with oarsmen,
load her with sacrificial beasts and put aboard
Khryseis in her loveliness. My deputy,
Aias, Idomeneus, or Prince Odysseus,
or you, Achilles, fearsome as you are,
will make the hecatomb and quiet the Archer."

Achilles frowned and looked at him, then said:

"You thick-skinned, shameless, greedy fool!
Can any Achaian care for you, or obey you,
after this on marches or in battle?
As for myself, when I came here to fight,
I had no quarrel with Troy or Trojan spearmen:
they never stole my cattle or my horses,
never in the black farmland of Phthia
ravaged my crops. How many miles there are
of shadowy mountains, foaming seas, between!
No, no, we joined for you, you insolent boor,
to please you, fighting for your brother's sake
and yours, to get revenge upon the Trojans.
You overlook this, dogface, or don't care,
and now in the end you threaten to take my girl,
a prize I sweated for, and soldiers gave me!

Never have I had plunder like your own
from any Trojan stronghold battered down
by the Achaians. I have seen more action
hand to hand in those assaults than you have,
but when the time for sharing comes, the greater
share is always yours. Worn out with battle

I carry off some trifle to my ships.
Well, this time I make sail for home.
Better to take now to my ships. Why linger,
cheated of winnings, to make wealth for you?"

To this the high commander made reply:

Desert, if that's the way the wind blows. Will I
beg you to stay on my account? I will not.
Others will honor me, and Zeus who views
the wide world most of all.

      No officer
is hateful to my sight as you are, none
given like you to faction, as to battle --
rugged you are, I grant, by some god's favor.
Sail, then, in your ships, and lord it over
your own battalion of Myrmidons. I do not
give a curse for you, or for your anger.
But here is warning for you:

      Khryseis
being required of me by Phoibos Apollo,
she will be sent back in a ship of mine,
manned by my people. That done, I myself
will call for Briseis at your hut, and take her,
flower of young girls that she is, your prize,
to show you here and now who is the stronger
and make the next man sick at heart -- if any
think of claiming equal place with me."

A pain like grief weighted on the son of Peleus,
and in his shaggy chest this way and that
the passion of his heart ran: should he draw
longsword from hip, stand off the rest, and kill
in single combat the great son of Atreus,
or hold his rage in check and give it time?
And as this tumult swayed him, as he slid
the big blade slowly from the sheath, Athena
came to him from the sky. The white-armed goddess,
Hera, sent her, being fond of both,
concerned for both men. And Athena, stepping
up behind him, visible to no one
except Achilles, gripped his red-gold hair.

Startled, he made a half turn, and he knew her
upon the instant for Athena: terribly

her grey eyes blazed at him. And speaking softly
but rapidly aside to her he said:

"What now, O daughter of the god of heaven
who bears the storm cloud, why are you here? To see
the wolfishness of Agamemnon?
Well, I give you my word: this time, and soon,
he pays for his behavior with his blood."

The grey-eyed goddess Athena said to him:

"It was to check this killing rage I came
from heaven, if you will listen. Hera sent me,
being fond of both of you, concerned for both.
Enough: break off this combat, stay your hand
upon the sword hilt. Let him have a lashing
with words, instead: tell him how things will be.
Here is my promise, and it will be kept:
winnings three times as rich, in due season,
you shall have in requital for his arrogance.
But hold your hand. Obey."

      The great runner,
Achilles, answered:
      "Nothing for it, goddess,
but when you two immortals speak, a man
complies, though his heart burst. Just as well.
Honor the gods' will, they may honor ours."
On this he stayed his massive hand
upon the silver pommel, and the blade
of his great weapon slid back in the scabbard.
The man had done her bidding. Off to Olympus,
gaining the air, she went to join the rest,
the powers of heaven in the home of Zeus.

But now the son of Peleus turned on Agamemnon
and lashed out at him, letting his anger ride
in execration:

      "Sack of wine,
you with your cur's eyes and your antelope heart!
You've never had the kidney to buckle on
armor among the troops, or make a sortie
with picked men -- oh, no; that way death might lie.
Safer, by god, in the middle of the army--
is it not? -- to commandeer the prize
of any man who stands up to you! Leech!

Commander of trash! If not, I swear,
you never could abuse one soldier more!

But here is what I say: my oath upon it
by this great staff: look: leaf or shoot
it cannot sprout again, once lopped away
from the log it left behind in the timbered hills;
it cannot flower, peeled of bark and leaves;
instead, Achaian officers in council
take it in hand by turns, when they observe
by the will of Zeus due order in debate:
let this be what I swear by then: I swear
a day will come when every Achaian soldier
will groan to have Achilles back. That day
you shall no more prevail on me than this
dry wood shall flourish -- driven though you are,
and though a thousand men perish before
the killer, Hector. You will eat your heart out,
raging with remorse for this dishonor
done by you to the bravest of Achaians."
He hurled the staff, studded with golden nails,
before him on the ground. Then down he sat,
and fury filled Agamemnon, looking across at him.
But for the sake of both men Nestor arose,
the Pylians' orator, eloquent and clear;
argument sweeter than honey rolled from his tongue.
By now he had outlived two generations
of mortal men, his own and the one after,
in Pylos land, and still ruled in the third.
In kind reproof he said:

        "A black day, this.
Bitter distress comes this way to Achaia.
How happy Priam and Priam's sons would be,
and all the Trojans -- wild with joy -- if they
got wind of all these fighting words between you,
foremost in council as you are, foremost
in battle. Give me your attention. Both
are younger men than I, and in my time
men who were even greater have I known
and none of them disdained me. Men like those
I have not seen again, nor shall: Peirithoos,
the Lord Marshal Dryas, Kaineus, Exadios,
Polyphemos, Theseus -- Aigeus' son,
a man like the immortal gods. I speak
of champions among men of earth, who fought
with champions, with wild things of the mountains,

great centaurs whom they broke and overpowered.
Among these men I say I had my place
when I sailed out of Pylos, my far country,
because they called for me. I fought
for my own hand among them. Not one man
alive now upon earth could stand against them.
And I repeat: they listened to my reasoning,
took my advice. Well, then, you take it too.
It is far better so.

       Lord Agamemnon,
do not deprive him of the girl, renounce her.
The army had allotted her to him.
Achilles, for your part , do not defy
your King and Captain. No one vies in honor
with him who holds authority from Zeus.
You have more prowess, for a goddess bore you;
his power over men surpasses yours.

But, Agamemnon, let your anger cool.
I beg you to relent, knowing Achilles
a sea wall for Achaians in the black waves of war."

Lord Agamemnon answered:
       "All you say is fairly said, sir, but this man's ambition,
remember, is to lead, to lord it over
everyone, hold power over everyone,
give orders to the rest of us! Well, one
will never take his orders! If the gods
who live forever made a spearman of him,
have they put insults on his lips as well?"

Achilles interrupted:
       "What a poltroon,
how lily-livered I should be called, if I
knuckled under to all you do or say!
Give your commands to someone else, not me!
And one more thing I have to tell you: think it
over: this time, for the girl, I will not
wrangle in arms with you or anyone,
though I am robbed of what was given me;
but as for any other thing I have
alongside my black ship, you shall not take it
against my will. Try it. Hear this, everyone:
that instant your hot blood blackens my spear!"

They quarreled in this way, face to face, and then
broke off the assembly by the ships. Achilles
made his way to his squadron and his quarter,
Patroklos by his side, with his companions.

Agamemnon proceeded to launch a ship,
assigned her twenty oarsmen, loaded beasts
for sacrifice to the god, then set aboard
Khryseis in her loveliness, The versatile
Odysseus took the deck, and, all oars manned,
they pulled out on the drenching ways of sea.
The troops meanwhile were ordered to police camp
and did so, throwing refuse in the water;
then to Apollo by the barren surf
they carried out full-tally hecatombs,
and the savor curled in crooked smoke toward heaven.

That was the day's work in the army.
        Agamemnon
had kept his threat in mind, and now he acted,
calling Eurybates and Talthybios,
his aides and criers:

        "Go along," he said,
"both of you, to the quarters of Achilles
and take his charming Briseis by the hand
to bring to me. And if he balks at giving her
I shall be there myself with men-at-arms
in force to take her -- all the more gall for him."
So, ominously, he sent them on their way,
and they who had no stomach for it went
along the waste sea shingle toward the ships
and shelters of the Myrmidons. Not far
from his black ship and hut they found the prince
in the open, seated. And seeing these two come
was cheerless to Achilles. Shamefast, pale
with fear of him, they stood without a word;
but he knew what they felt and called out:

        "Peace to you,
criers and couriers of Zeus and men!
Come forward. Not one thing have I against you:
Agamemnon is the man who sent you
for Briseis. Here then, my lord Patroklos,
bring out the girl and give her to these men.
And let them both bear witness before the gods
who live in bliss, as before men who die,

including this harsh king, if ever hereafter
a need for me arises to keep the rest
from black defeat and ruin.

      Lost in folly,
the man cannot think back or think ahead
how to come through a battle by the ships."
Patroklos did the bidding of his friend,
led from the hut Briseis in her beauty
and gave her to them. Back along the ships
they took their way, and the girl went, loath to go.

Leaving his friends in haste, Achilles wept,
and sat apart by the grey wave, scanning the endless sea.
Often he spread his hands in prayer to his mother:

"As my life came from you, though it is brief,
honor at least from Zeus who storms in heaven
I call my due. He gives me precious little.
See how the lord of the great plains, Agamemnon,
humiliated me! He has my prize,
by his own whim, for himself."

      Eyes wet with tears,
he spoke, and her ladyship his mother heard him
in green deeps where she lolled near her old father.
Gliding she rose and broke like mist from the inshore
grey sea face, to sit down softly before him,
her son in tears; and fondling him she said:

"Child, why do you weep? What grief is this?
Out with it, tell me, both of us should know."
Achilles, fast in battle as a lion,
groaned and said:

      "Why tell you what you know?
We sailed out raiding, and we took by storm
that ancient town of Eetion called Thebe,
plundered the place, brought slaves and spoils away.
at the division, later,
they chose a young girl, Khryseis, for the king.
Then Khryses, priest of the Archer God, Apollo,
came to the beachhead we Achaians hold,
bringing no end of ransom for his daughter;
he had the god's white bands on a golden staff
and sued for grace from the army of Achaia,
mostly the two Atreidai, corps commanders.

All of our soldiers murmured in assent:
'Behave well to the priest. And take the ransom!'
But Agamemnon would not. It went against his desire,
and brutally he ordered the man away.
So the old man withdrew in grief and anger.
Apollo cared for him: he heard his prayer
and let black bolts of plague fly on the Argives.

One by one our men came down with it
and died hard as the god's shots raked the army
broadside. But our priest divined the cause
and told us what the god meant by the plague.

I said, 'Appease the god!' but Agamemnon
could not contain his rage; he threatened me,
and what he threatened is now done--
one girl the Achaians are embarking now
for Khryse beach with gifts for Lord Apollo;
the other, just now, from my hut -- the criers
came and took her, Briseus' girl, my prize,
given by the army.

        If you can, stand by me:
go to Olympus, pray to Zeus, if ever
by word or deed you served him--
and so you did, I often heard you tell it
in Father's house: that time when you alone
of all the gods shielded the son of Kronos
from peril and disgrace--when other gods,
Pallas, Athena, Hera, and Poseidon,
wished him in irons, wished to keep him bound,
you had the will to free him of that bondage,
and called up to Olympus in all haste
Aigaion, whom the gods call Briareus,
the giant with a hundred arms, more powerful
than the sea-god, his father. Down he sat
by the son of Kronos, glorying in that place.
For fear of him the blissful gods forbore
to manacle Zeus.

        Remind him of these things,
cling to his knees and tell him your good pleasure
if he will take the Trojan side
and roll the Achaians back to the water's edge,
back on the ships with slaughter! All the troops
may savor what their king has won for them,
and he may know his madness, what he lost

when he dishonored me, peerless among Achaians."
Her eyes filled, and a tear fell as she answered:

"Alas, my child, why did I rear you, doomed
the day I bore you? Ah, could you only be
serene upon this beachhead through the siege,
your life runs out so soon.
Oh early death! Oh broken heart! No destiny
so cruel! And I bore you to this evil!

But what you wish I will propose
To Zeus, lord of the lightning, going up
myself into the snow-glare of Olympus
with hope for his consent.

        Be quiet now
beside the long ships, keep your anger bright
against the army, quit the war.
        Last night
Zeus made a journey to the shore of Ocean
to feast among the Sunburned, and the gods
accompanied him. In twelve days he will come
back to Olympus. Then I shall be there
to cross his bronze doorsill and take his knees.
I trust I'll move him."

        Thetis left her son
still burning for the softly belted girl
whom they had wrested from him.
        Meanwhile Odysseus
with his shipload of offerings came to Khryse.
Entering the deep harbor there
they furled the sails and stowed them, and unbent
forestays to ease the mast down quickly aft
into its rest; then rowed her to a mooring.
Bow-stones were dropped, and they tied up astern,
and all stepped out into the wash and ebb,
then disembarked their cattle for the Archer,
and Khryseis, from the deep sea ship. Odysseus,
the great tactician, led her to the altar,
putting her in her father's hands, and said:

"Khryses, as Agamemnon's emissary
I bring your child to you, and for Apollo
a hecatomb in the Danaans' name.
We trust in this way to appease your lord,
who sent down pain and sorrow on the Argives."

So he delivered her, and the priest received her,
the child so dear to him, in joy. Then hastening
to give the god his hecatomb, they led
bullocks to crowd around the compact altar,
rinsed their hands and delved in barley baskets,
as open-armed to heaven Khryses prayed:

"Oh hear me, master of the silver bow,
protector of Tenedos and the holy towns,
if while I prayed you listened once before
and honored me, and punished the Achaians,
now let my wish come true again. But turn
your plague away this time from the Danaans."

And this petition, too, Apollo heard.
When prayers were said and grains of barley strewn,
they held the bullocks for the knife, and flayed them,
cutting out joints and wrapping these in fat,
two layers, folded, with raw strips of flesh,
for the old man to burn on cloven faggots,
wetting it all with wine.

        Around him stood
young men with five-tined forks in hand, and when
the vitals had been tasted, joints consumed,
they sliced the chines and quarters for the spits,
roasted them evenly and drew them off.
Their meal being now prepared and all work done,
they feasted to their hearts' content and made
desire for meat and drink recede again,
then young men filled their wine bowls to the brim,
ladling drops for the god in every cup.
Propitiatory songs rose clear and strong
until day's end, to praise the god, Apollo,
as One Who Keeps the Plague Afar; and listening
the god took joy.

        After the sun went down
and darkness came, at last Odysseus' men
lay down to rest under the stern hawsers.

When Dawn spread out her fingertips of rose
they put to sea for the main camp of the Achaians,
and the Archer God sent them a following wind.
Stepping the mast they shook their canvas out,
and wind caught, bellying the sail. A foaming
dark blue wave sang backward from the bow

as the running ship made way against the sea,
until they came offshore of the encampment.
Here they put in and hauled the black ship high,
far up the sand, braced her with shoring timbers,
and then disbanded, each to his own hut.

Meanwhile unstirring and with smoldering heart,
the godlike athlete, son of Peleus, Prince
Achilles waited by his racing ships.
He would not enter the assembly
of emulous men, nor ever go to war.
but felt his valor stalling in his breast
with idleness, and missed the cries of battle.

Now when in fact twelve days had passed, the gods
who live forever turned back to Olympus,
with Zeus in power supreme among them.
    Thetis
had kept in mind her mission for her son,
and rising like a dawn mist from the sea
into a cloud she soared aloft in heaven
to high Olympus. Zeus with massive brows
she found apart, on the chief crest enthroned,
and slipping down before him, her left hand
placed on his knees and her right hand held up
to cup his chin, she made her plea to him:

"O Father Zeus, if ever amid immortals
by word or deed I served you, grant my wish
and see to my son's honor! Doom for him
of all men comes on quickest.
    Now Lord Marshal
Agamemnon has been highhanded with him,
has commandeered and holds his prize of war.
But you can make him pay for this, profound
mind of Olympus!
    Lend the Trojans power,
until the Achaians recompense my son
and heap new honor upon him!"

    When she finished,
the gatherer of cloud said never a word
but sat unmoving for a long time, silent.
Thetis clung to his knees, then spoke again:

"Give your infallible word, and bow your head,
or else reject me. Can you be afraid

to let me see how low in your esteem
I am of all the gods?"

      Greatly perturbed,
Lord Zeus who masses cloud said:
      "Here is trouble.
You drive me into open war with Hera
sooner or later:
she will be at me, scolding all day long.
Even as matters stand she never rests
from badgering me before the gods: I take
the Trojan side in battle, so she says.

Go home before you are seen. But you can trust me
to put my mind on this; I shall arrange it.
Here let me bow my head, then be content
to see me bound by that most solemn act
before the gods. My word is not revocable
nor ineffectual, once I nod upon it."
He bent his ponderous black brows down, and locks
ambrosial of his immortal head
swung over them, as all Olympus trembled.
After this pact they parted: misty Thetis
from glittering Olympus leapt away
into the deep sea; Zeus to his hall retired.
There all the gods rose from their seats in deference
before their father; not one dared
face him unmoved, but all stood up before him,
and thus he took his throne.

      But Hera knew
he had new interests; she had seen
the goddess Thetis, silvery-footed daughter
of the Old One of the sea, conferring with him,
and, nagging, she inquired of Zeus Kronion:

"Who is it this time, schemer? Who has your ear?
How fond you are of secret plans, of taking
decisions privately! You could not bring yourself,
could you, to favor me with any word
of your new plot?"

      The father of gods and men
said in reply:
      Hera, all my provisions
you must not itch to know.
You'll find them rigorous, consort though you are.

In all appropriate matters no one else,
no god or man, shall be advised before you.
But when I choose to think alone,
don't harry me about it with your questions."
The Lady Hera answered, with wide eyes:

"Majesty, what a thing to say. I have not
'harried' you before with questions, surely;
you are quite free to tell what you will tell.
This time I dreadfully fear -- I have a feeling--
Thetis, the silvery-footed daughter
of the Old One of the sea, led you astray.
Just now at daybreak, anyway, she came
to sit with you and take your knees; my guess is
you bowed your head for her in solemn pact
that you will see to the honor of Achilles --
that is, to Achaian carnage near the ships."

Now Zeus the gatherer of cloud said:

        "Marvelous,
you and your guesses; you are near it, too.
But there is not one thing that you can do about it,
only estrange yourself still more from me--
all the more gall for you. If what you say
is true, you may be sure it pleases me.
And now you just sit down, be still, obey me,
or else not all the gods upon Olympus
can help in the least when I approach your chair
to lay my inexorable hands upon you."
At this the wide-eyed Lady Hera feared him,
and sat quite still, and bent her will to his.
Up through the hall of Zeus now all the lords
of heaven were sullen and looked askance.
Hephaistos, master artificer, broke the silence,
doing a kindness to the snowy-armed
lady, his mother Hera.

        He began:
"Ah, what a miserable day, if you two
raise your voices over mortal creatures!
More than enough already! Must you bring
your noisy bickering among the gods?
What pleasure can we take in a fine dinner
when baser matters gain the upper hand?
To Mother my advice is -- what she knows
better make up to Father, or he'll start

his thundering and shake our feast to bits.
You know how he can shock us if he cares to--
out of our seats with lightning bolts!
Supreme power is his. Oh, soothe him, please,
take a soft tone, get back in his good graces.
Then he'll be benign to us again."
He lurched up as he spoke, and held up a wine cup
out to her, a double-handed one,
and said:

       "Dear Mother, patience, hold your tongue,
no matter how upset you are. I would not
see you battered, dearest.

       It would hurt me,
and yet I could not help you, not a bit.
The Olympian is difficult to oppose.
One other time I took your part he caught me
around one foot and flung me
into the sky from our tremendous terrace.
I soared all day! Just as the sun dropped down
I dropped down, too, on Lemnos -- nearly dead.
The island people nursed a fallen god."

He made her smile--and the goddess, white-armed Hera,
smiling took the wine cup from his hand.
Then, dipping from the wine bowl, round he went
from left to right, serving the other gods
nectar of sweet delight.

       And quenchless laughter
broke out among the blissful gods
to see Hephaistos wheezing down the hall.
So all day long until the sun went down
they spent in feasting, and the measured feast
matched well their hearts' desire.
So did the flawless harp held by Apollo
and heavenly songs in choiring antiphon
that all the Muses sang.

       And when the shining
sun of day sank in the west, they turned
homeward each one to rest, each to that home
the bandy-legged wondrous artisan
Hephaistos fashioned for them with his craft.
The lord of storm and lightning, Zeus, retired
and shut his eyes where sweet sleep ever came to him,
and at his side lay Hera, Goddess of the Golden Chair.

Book Two - Assembly and Muster of Armies

Now slept the gods, and those who fought at Troy--
horse-handlers, charioteers -- the long night through,
but slumber had no power over Zeus,
who pondered in the night how to exalt
Achilles, how in his absence to destroy
Achaians in windrows at the ships.
He thought it best to send to Agamemnon
that same night a fatal dream.
Calling the dream he said:
            "Sinister Dream,
go down amid the fast ships of Achaia,
enter Lord Agamemnon's quarters, tell him
everything, point by point, as I command you:
Let him prepare the long-haired carls of Achaia
to fight at once. Now he may take by storm
the spacious town of Troy. The Olympians, tell him,
are of two minds no longer: Hera swayed them,
and black days overhang the men of Troy."

The dream departed at his word, descending
swift as wind to where the long ships lay,
and sought the son of Atreus. In his hut he found him sleeping,
drifted all about with balm of slumber. At the marshal's pillow
standing still, the dream took shape
as Neleus' son, old Nestor. Agamemnon
deferred to Nestor most, of all his peers;
so in his guise the dream spoke to the dreamer:

"Sleeping, son of Atreus, tamer of horses?
You should not sleep all night, not as a captain
responsible for his men, with many duties,
a great voice in the conferences of war.
Follow me closely: I am a messenger
from Zeus, who is far away but holds you dear.
'Prepare the troops,' he said, 'to take the field
without delay: now may you take by storm
the spacious town of Troy. The Olympian gods
are of two minds no longer: Hera's pleading
swayed them all, and bitter days from Zeus
await the Trojans.' Hold on to this message
against forgetfulness in tides of day
when blissful sleep is gone."

On this the dream
withdrew into the night, and left the man
to envision, rapt, all that was not to be,
thinking that day to conquer Priam's town.
Oh childish trust! What action lay ahead
in the mind of Zeus he could not know -- what grief
and wounds from shock of combat in the field,
alike for Trojans and Achaians.

Waking,
he heard the dream voice ringing round him still,
and sat up straight to pull his tunic on,
a fresh one, never worn before. He shook
his cloak around him, tied his shining feet
in fitted sandals, hung upon his shoulder
baldric and long sword, hilted all in silver,
and, taking his dynastic staff in hand,
he made his way among the ships.

Pure Dawn
had reached Olympus' mighty side,
heralding day for Zeus and all the gods,
as Agamemnon, the Lord Marshal, met
his clarion criers and directed them
to call the unshorn Achaians to full assembly.
The call sang out, and quickly they assembled.
But first, alongside Nestor's ship, he held
a council with his peers -- there he convened them
and put a subtle plan before them, saying:

"Hear me, friends. A vision in a dream
has come to me in the starry night -- a figure
in height and bearing very close to Nestor,
standing above my pillow, saying to me:
'Sleeping, son of Atreus, tamer of horses?
You should not sleep all night, not as a captain
responsible for his men, with many duties,
a great voice in the conferences of war.
Follow me closely: I am a messenger
from Zeus, who is far away but holds you dear.
"Prepare the troops," he said, "to take the field
without delay: now may you take by storm
the spacious town of Troy. The Olympian gods
are of two minds no longer: Hera's pleading
swayed them all, and bitter days from Zeus
await the Trojans." Hold on to this message.'
When he had said all this, the phantasm

departed like a bird, and slumber left me.
Look to it then, we arm the troops for action --
but let me test them first, in that harangue
that custom calls for. What I shall propose
is flight in the long ships! You must hold them back,
speaking each one from where he stands."
      How curtly
he told his curious plan, and took his seat!
Now stood Lord Nestor of the sandy shore
of Pylos, in concern for them, and spoke:

"Friends, lord and captains of the Argives,
if any other man had told this dream,
a fiction, we should call it; we'd be wary.
But he who saw the vision is our king.
Up with you, and we'll put the men in arms."

On this he turned and led the way from council,
and all the rest, staff-bearing counselors,
rose and obeyed their marshal.
      From the camp
the troops were turning out now, thick as bees
that issue from some crevice in a rock face,
endlessly pouring forth, to make a cluster
and swarm on blooms of summer here and there,
glinting and droning, busy in bright air.
Like bees innumerable from ships and huts
down the deep foreshore streamed those regiments
toward the assembly ground--and Rumor blazed
among them like a crier sent from Zeus.
Turmoil grew in the great field as they entered
and sat down, clangorous companies, the ground
under them groaning, hubbub everywhere.
Now nine men, criers, shouted to compose them:

"Quiet! Quiet! Attention! Hear our captains!"

Then all strove to their seats and hushed their din.
Before them now arose Lord Agamemnon,
holding the staff Hephaistos fashioned once
and took pains fashioning: it was a gift
from his to the son Kronos, lordly Zeus,
who gave it to the bright pathfinder, Hermes.
Hermes handed it on in turn to Pelops,
famous charioteer, Pelops to Atreus,
and Atreus gave it to the sheepherder
Thyestes, he to Agamemnon, king

and lord of many islands, of all Argos-
the very same who leaning on it now
spoke out among the Argives:

"Friends, fighters, Danaans, companions of Ares,
the son of Kronos has entangled me
in cruel folly, wayward god! He promised
solemnly that I should not sail
before I stormed the inner town of Troy.
Crookedness and duplicity, it is clear!
He calls me to return to Argos beaten,
after these many losses.

                    That must be
his will and his good pleasure, who knows why?
Many a great town's height has he destroyed
and will destroy, being supreme in power.
Shameful indeed that future men should hear
we fought so long here, with such weight of arms,
all uselessly! We made long war for nothing,
never an end to it, though we had the odds.
The odds -- if we Achaians and the Trojans
should hold a truce and tally on both sides,
on one side native Trojans, on the other
Achaian's troops drawn up in squads of ten,
and each squad took one Trojan for a steward,
then many squads would go unserved. I tell you,
Achaian men so far outnumber those
whose home is Troy!"

                    But the allies are there.
From many Asian cities came these lances,
and it is they who hedge me out and hinder me
from plundering the fortress town of Troy.
Under great Zeus nine years have passed away,
making ship timbers rot, old tackle fray,
while overseas our wives and children still
await us in our halls. And yet the mission
on which we came is far from being done.
Well and good; let us act on what I say:
Retreat! Embark for our own fatherland!
We cannot hope any longer to take Troy!

He made their hearts leap in their breast, the rank
and file, who had no warning of his plan,
and all that throng, aroused, began to surge
as ground swells do on dark Ikarian deeps

under the south and east wind heeling down
from Father Zeus's cloudland--

      or a field
of standing grain when wind-puffs from the west
cross it in billows, and the tasseled ears
are bent and tossed: just so moved this assembly.
Shouting confusedly, they all began
to scramble for the ships. High in the air
a dust cloud from their scuffling rose, commands
rang back and forth -- to man the cables, haul
the black ships to the salt immortal sea.
They cleared the launching ways, their hearts on home,
and shouts went up as props were pulled away.

Thus, overriding their own destiny,
the Argives might have had their voyage homeward,
had Hera not resorted to Athena
and cried:

      "Can you believe it? Tireless
daughter of Zeus who bears the shield of cloud,
will they put out for home this way, the Argives,
embarking on the broad back of the sea!
How could they now abandon Helen,
princess of Argos -- leave her in Priam's hands,
the boast of every Trojan? Helen, for whom
Achaians died by thousands far from home?
Ah, go down through the ranks of men-at-arms;
in your mild way dissuade them, one by one,
from hauling out their graceful ships to sea!"

The grey-eyed goddess Athena obeyed her, diving
swifter than wind, down from the crest of Olympus,
to earth amid the long ships. There she found
Odysseus, peer of Zeus in stratagems,
holding his ground.

      He had not touched the prow
of his black ship, not he, for anguish filled him,
heart and soul; and halting near him now,
the grey-eyed goddess made her plea to him:
"Son of Laertes and the gods of old,
Odysseus, master mariner and soldier,
must all of you take oars in the long ships
in flight to your old country? Leaving Helen
in Priam's hands -- that Argive grace, to be

the boast of every Trojan? Helen, for whom
Achaians died by thousands, far from home?
No, no take heart, and go among the men;
in your mild way dissuade them, one by one,
from hauling out their graceful ships to sea."

Knowing the goddess' clear word when he heard it,
Odysseus broke into a run. He tossed
his cloak to be picked up by his lieutenant,
Eurybates of Ithaca, and wheeling
close to the silent figure of Agamemnon
relieved him of his great dynastic staff,
then ran on toward the ships.
        Each time he met
an officer or man of rank he paused
and in his ear he said:
        "Don't be a fool!
It isn't like you to desert the field
the way some coward would! Come, halt, command
the troops back to their seats. You don't yet know
what Agamemnon means.
        He means to test us,
and something punitive comes next. Not everyone
could hear what he proposed just now in council.
Heaven forbid he cripple, in his rage,
the army he commands. There's passion in kings;
they hold power from Zeus, they are dear to Zeus!"

But when Odysseus met some common soldier
bawling still, he drove him back; he swung
upon him with his staff and told him:
        "Fool,
go back, sit down, listen to better men --
unfit for soldiering as you are, weak sister,
counting for nothing in battle or in council!
Shall we all wield the power of kings? We can not,
and many masters are no good at all.
Let there be one commander, one authority,
holding his royal staff and precedence
from Zeus, the son of crooked-minded Kronos:
one to command the rest."
        So he himself
in his commanding way went through the army,
and back to the assembly ground they streamed
from ships and huts with multitudinous roar,
as when a comber from the windy sea
on a majestic beach goes thundering down,

and the ebb seethes offshore.
      So all subsided,
except one man who still railed on alone --
Thersites, a blabbing soldier,
who had an impudent way with officers,
thinking himself amusing to the troops--
the most obnoxious rogue who went to Troy.
Bowlegged, with one limping leg, and shoulders
rounded above his chest, he had a skull
quite conical, and mangy fuzz like mold.
Odious to Achilles this man was,
and to Odysseus, having yapped at both,
but this time he berated Agamemnon--
at whom in fact the troops were furious --
lifting his voice and jeering:
      "Agamemnon!
What have you got to groan about? What more
can you gape after? Bronze fills all your huts,
bronze and the hottest girls -- we hand them over
to you, you first, when any stronghold falls.
Or is it gold you lack? A Trojan father
will bring you gold in ransom for his boy --
though I -- or some foot soldier like myself --
roped the prisoner in.
      Or a new woman
to lie with, couple with, keep stowed away
for private use -- is that your heart's desire?
You send us back to bloody war for that?
Comrades! Are you women of Achaia?
I say we pull away for home, and leave him
here on the beach to lay his captive girls!
Let him find out if we troops are dispensable
when he loses us! Contempt is all he shows
for a man twice his quality, by keeping
Achilles' woman that he snatched away.
But there's no bile, no bad blood in Achilles;
he lets it go.
      Sir, if he drew his blade,
you'd never abuse another man!"

      So boldly
Thersites baited Marshal Agamemnon,
till at his side, abruptly,
Odysseus halted, glaring, and grimly said:

"You spellbinder! You sack of wind! Be still!
Will you stand up to officers alone?

Of all who came here to beleaguer Troy
I say there is no soldier worse than you.
Better not raise your voice to your commanders,
or rail at them, after you lie awake
with nothing on your mind but shipping home.
We have no notion, none, how this campaign
may yet turn out. Who knows if we sail homeward
in victory or defeat? Yet you bleat on,
defaming the Lord Marshal Agamemnon
because our Danaan veterans award him
plentiful gifts of war. You sicken me!

Here is my promise, and it will be kept:
if once again I hear your whining voice,
I hope Odysseus' head may be knocked loose
from his own shoulder, hope I may no longer
 be called the father of Telemakhos,
if I do not take hold of you and strip you --
yes, even of the shirt that hides your scut!
From this assembly ground I'll drive you howling
and whip you like a dog into the ships!"
At this he struck him sharply with his staff
on ribs and shoulders. The poor devil quailed,
and a welling tear fell from his eyes. A scarlet
welt, raised by the golden-studded staff,
sprang out upon his back. Then, cowering down
in fear and pain he blinked like an imbecile
and wiped his tears upon his arms.
　　　　The soldiers,
for all their irritation, fell to laughing
at the man's disarray. You might have heard
one fellow, glancing at his neighbor, say:

"Oh, what a clout! A thousand times Odysseus
has done good work, thinking out ways to fight
or showing how you do it: this time, though,
he's done the best deed of the war,
making that poisonous clown capsize. By god,
a long, long time will pass before our hero
cares to call down his chief again!"

　　　　The crowd
took it in this way. But the raider of cities,
Odysseus, with his staff, stood upright there,
and at his side grey-eyed Athena stood
in aspect like a crier, calling: "Silence!"
that every man, front rank and rear alike,

might hear his words and weigh what he proposed.
Now for their sake he spoke:
            "Lord Agamemnon,
son of Atreus, king, your troops are willing
to let you seem disgraced in all men's eyes;
they will not carry through the work they swore to
en route from Argos, from the bluegrass land,
never to turn back till they plundered Troy.
No, now like callow boys or widowed women
they wail to one another to go home!

I grant this hardship wearying to everyone.
I grant the urge to go. Who can forget,
one month at sea -- no more -- far from his wife
will make a raider sick of the rowing bench,
sick of his ship, as gales and rising seas
delay him, even a month! As for ourselves,
now is the ninth year that we keep the siege.
No wonder at it, then: I cannot blame
you men for sickening by the beaked ships!
Ah, but still it would be utter shame
to stay so long and sail home empty-handed.
Hold on hard, dear friends!
Come, sweat it out, until at least we learn
if Kalchas made true prophecy or not.
Here is a thing we cannot help remembering,
and every man of you whom death has spared
can testify:
            One day, just when the ships
had staged at Aulis, loaded, every one,
with woe for Priam and the men of Troy,
we gathered round a fountain by the altars,
performing sacrifices to the gods
under a dappled sycamore. The water
welled up shining there, and in that place
the great portent appeared.
            A blood-red serpent
whom Zeus himself sent gliding to the light,
blood-chilling, silent, from beneath our altar,
twined and swiftly spiraled up the tree.
There were some fledgling sparrows, baby things,
hunched in their downy wings -- just eight of these
among the leaves along the topmost bough,
and a ninth bird, the mother who had hatched them.
The serpent slid to the babies and devoured them,
all cheeping pitifully, while their mother
fluttered and shrilled in her distress. He coiled

and sprang to catch her by one frantic wing.

After the snake had gorged upon them all,
the god who sent him turned him to an omen:
turned him to stone, hid him in stone- -- a wonder
worked by the son of crooked-minded Kronos --
and we stood awed by what had come to pass.
Seeing this portent of the gods had visited
our sacrifices, Kalchas told the meaning
before us all, at once. He said:

'Dumbfounded, are you, gentlemen of Achaia?
Here was a portent for us, and a great one,
granted us by inscrutable Zeus -- a promise
long to be in fulfillment -- but the fame
of that event will never die.
        Consider:
the snake devoured fledglings and their mother,
the little ones were eight, and she made nine.
Nine are the years that we shall wage this war,
and in the tenth we'll take the spacious town.'

That was his explanation of the sign.
Oh, see now, how it all comes true! Hold out,
Achaians with your gear of war, campaigners,
hold on the beachhead till we take the town!"

After the speech a great shout from the Argives
echoed fiercely among the ships: they cried
"Aye" to noble Odysseus' words. Then Nestor,
lord of Gerenia, charioteer, addressed them:

"Lamentable, the way you men have talked,
like boys, like children, strangers to stern war.
What will become of our sworn oaths and pacts?
"To the flames,' you mean to say, 'with battle plans,
soldierly calculations, covenants
our right hands pledged and pledged with unmixed wine'?
Once we could trust in these. But wrangling now
and high words dissipate them, and we cannot
turn up a remedy, though we talk for days.

Son of Atreus, be as you were before,
inflexible; commit the troops to combat;
let those go rot, those few, who take their counsel
apart from the Achaians. They can win nothing by it.
They would sail for Argos before they know if what the

Lord Zeus promised will be proved false or true.
I think myself the power above us nodded
on that day when the Argives put to sea
in their fast sailing ships, with death aboard
and doom for Trojans.
        Forking out of heaven,
he lightened on the right -- a fateful sign.
Therefore let no man press for our return
before he beds down with some Trojan wife,
to avenge the struggles and the groans of Helen.

If any man would sooner die than stay,
let him lay hand upon his ship--
he meets his death and doom before the rest.
My lord, yourself be otherwise persuaded.
What I am going to say is not a trifle
to toss aside: marshal the troops by nations
and then again by clans, Lord Agamemnon,
clan in support of clan, nation of nation.
If you will do this, and they carry it out,
you may find out which captains are poltroon
and which are valorous; foot soldiers, too;
as each will fight before his clansmen's eyes
when clans make up our units in the battle.
You can discern then if your siege has failed
by heaven's will or men's faintheartedness
and foolishness in war."
        Lord Agamemnon made reply:
        "Believe me, sir, once more
you win us all with your proposals here.
O Father Zeus, Athena, and Apollo,
give me ten more to plan with me like this
among the Achaians! Priam's fortress then
falls in a day, our own hands' prey and spoil.
But Zeus the storm king sent me misery,
plunging me into futile brawls and feuds.
I mean Achilles and myself. We fought
like enemies, in words, over a girl --
and I gave way to anger first.
        We two --
if we could ever think as one, the Trojans'
evil day would be postponed no longer.

Take your meal, now; we prepare for combat.
Let every man be sure his point is whetted,
his shield well slung. Let every charioteer
give fodder to his battle team, inspect

his wheels and car, and put his mind on war,
so we may bear ourselves as men all day
in the grim battle. There will be no respite,
no break at all -- unless night coming on
dissolves the battle lines and rage of men.
The shield strap will be sweat-soaked on your ribs,
your hand will ache and stiffen on the spear shaft,
and sweat will drench the horses' flanks that toil
to pull your polished car.

      But let me see
one man of you willing to drop out -- one man
skulking around the ships, and from that instant
he has no chance against the dogs and kites!"

Being so dismissed, the Argives roared, as when
upon some cape a sea roused by the south wind
roars on a jutting point of rock, a target
winds and waves will never let alone,
from any quarter rising. So the soldiers
got to their feet and scattered to the ships
to send up smoke from campfires and be fed.
But first, to one of the gods who never die,
each man resigned his bit and made his prayer
to keep away from death in that day's fighting.
As for Lord Agamemnon, their commander,
a fattened ox he chose for sacrifice
to Zeus the overlord of heaven -- calling round him
the senior captains of the Achaian host:
Nestor, then Lord Idomeneus, then those
two lords who bore the name of Aias, then
the son of Tydeus and, sixth, Odysseus,
the peer of Zeus in war craft. Menelaos,
lord of the war cry, needed no summoning;
he knew and shared the duties of his brother.
Around the ox they stood, and took up barley,
and Agamemnon prayed on their behalf:

"O excellency, O majesty, O Zeus
beyond the storm cloud, dwelling in high air,
let not the sun go down upon this day
into the western gloom, before I tumble
Priam's blackened rooftree down, exploding
fire through his portals! Let me rip
with my bronze point the shirt that clings on Hektor
and slash his ribs! May throngs around him lie --
his friends, head-down in dust, biting dry ground!"

But Zeus would not accomplish these desires.
He took the ox, but added woe on woe.

When prayers were said and grains of barley strewn,
they held the bullock for the knife and flayed him,
cutting out joints and wrapping these in fat,
two layers, folded, with raw strips of flesh,
to burn on cloven faggots, and the tripes
they spitted to be broiled. When every joint
had been consumed, and kidneys had been tasted,
they sliced the chines and quarters for the spits,
roasted them evenly and drew them off.
The meal being now prepared and all work done,
they feasted royally and put away
desire for meat and drink.

Then Nestor spoke:
"Excellency, Lord Marshal Agamemnon,
we shall do well to tarry here no longer,
we officers, in our circle. Let us not
postpone the work heaven put into our hands.
Let criers among the Achaian men-at-arms
muster our troops along the ships. Ourselves,
we'll pass together down the Achaian lines
to rouse their appetite for war."

And Agamemnon,
marshal of the army, turned at once,
telling his criers to send out shrill and clear
to all Achaian troops the call to battle.
The cry went out, the men came crowding, officers
from their commander's side went swiftly down
to form each unit -- and the grey-eyed goddess
Athena kept the pace behind them, bearing
her shield of storm, immortal and august,
whose hundred golden-plaited tassels, worth
a hecatomb each one, floated in air.
So down the ranks that dazzling goddess went
to stir the attack, and each man in his heart
grew strong to fight and never quit the melee.
for at her passage war itself became
lovelier than return, lovelier than sailing
in the decked ships to their own native land.

As in dark forests, measureless along
the crest of hills, a conflagration soars,
and the bright bed of fire glows for miles,

now fiery lights from this great host in bronze
played on the earth and flashed high into heaven.

And as migrating birds, nation by nation,
wild geese and arrow-throated cranes and swans,
over Asia's meadowland and marches
around the streams of Kaystrios, with giant
flight and glorying wings keep beating down
in tumult on that verdant land
that echoes to their pinions, even so,
nation by nation, from the ships and huts,
this host debouched upon Skamander plain.
With noise like thunder pent in earth
under their trampling, under the horses' hooves,
they filled the flowering land beside Skamander,
as countless as the leaves and blades of spring.
So, too, like clouds of buzzing, fevered flies
that swarm about a cattle stall in summer
when pails are splashed with milk: so restlessly
by thousands moved the fighters of Achaia
over the plain, lusting to rend the Trojans.
But just as herdsmen easily divide
their goats when herds have mingled in a pasture,
so these were marshaled by their officers
to one side and the other, forming companies
for combat.
      Agamemnon's lordly mien,
was like the mien of Zeus whose joy is lightning;
oaken-waisted as Ares, god of war,
he seemed , and deep-chested as Lord Poseidon,
and as a great bull in his majesty
towers supreme amid a grazing herd,
so on that day Zeus made the son of Atreus
tower over his host, supreme among them.
Tell me now, Muses, dwelling on Olympus,
as you are heavenly, and are everywhere,
and everything is known to you -- while we
can only hear the tales and never know --
who were the Danaan lords and officers?
The rank and file I shall not name; I could not,
if I were gifted with ten tongues and voices
unfaltering, and a brazen heart within me,
unless the Muses, daughters of Olympian
Zeus beyond the storm cloud, could recall
all those who sailed for the campaign of Troy.
Let me name only captains of contingents
and number all the ships.

        Of the Boiotians,
Penelos, Leitos, Arkesilaos,
Prothoenor, and Klonios were captains.
Boiotians -- men Hyria and Aulis,
the stony town, and those who lived at Skhoinos
and Skolos and the glens of Eteonos;
Thespeia; Graia, round the dancing grounds
of Mykalessos; round the walls of Harma,
Eilesion, Erythrai, Eleon,
Hyle and Peteon, and Okalea,
and Medeon, that compact citadel,
Kopai, Eutresis, Thiesbe of the doves;
those, too, of Koroneia, and the grassland
of Haliartos, and the men who held
Plataia town and Glisas, and the people
of Lower Thebes, the city ringed with walls,
and great Ongkhestos where Poseidon's grove
glitters; and people, too, of Arne, rich
in purple wine grapes, and the men of Mideia,
Nisa the blest, and coastal Anthedon.
All these had fifty ships. One hundred twenty
Boiotain fighters came in every ship.

Their neighbors of Asplendon, then, Minyan
Orkhomenos, Askalaphos their captain
with Ialmenos, both sons of Ares, both
conceived in Aktor's manor by severe
Astyokhe, who kept a tryst with Ares
in the women's rooms above, where secretly
the strong god lay beside her. Thirty ships
these Minyans drew up in line of battle.

Phokians in their turn were led by Skhedios
and by Epistrophos, and sons of Iphitos
Naubolides, that hero; Phokians
dwelling in Kyparissos, rocky Pytho,
Krisa the holy, Panopeus and Daulis,
near Anemoreia, Hyampolis,
and by the side of noble Kephisos,
or in Lilaia, where that river rises.
Forty black ships had crossed the sea with these,
who now drew up their companies on the flank
of the Boiotians, and armed themselves.

The Lokrians had Aias for commander,
Oileus' son, that Aias known as the Short One
as being neither tall nor great compared

with Aias Telamonios. A corselet
all of linen he wore, and could out throw
all Hennenes and Achaians with a spear.
His were the Lokrians who lived at Kynos.
and Opoeis and Kalliaros,
Bessa and Skarphe and the pretty town
Augeiai; Tarphe and Thronion that lie
on both sides of the stream Boagrios.
Aias led forty black ships of the Lokrians
who lived across the channel from Euboia.
Men of that island, then, the resolute
Abantes, those of Khalkis, Eiretrice,
and Histiaia, of the laden vines,
Kerinthos by the sea, the crag of Dion,
those of Karystos, those of Styra -- all
who had young Elephenor Khalkodontiades,
the chief of the Abantes, for commander.
Quick on their feet, with long scalp locks, those troops
enlisted hungering for body armor
of enemies to pierce with ashen spears;
and Elephenor's black ships numbered forty.

Next were the men of Athens, that strong city,
the commonwealth protected by Erekhtheus.
He it was whom Athena, Zeus's daughter,
cared for in childhood in the olden time --
though he was born of plow land kind with grain.
She placed him in her city, in her shrine,
where he receives each year, with bulls and rams,
the prayers of young Athenians. Their commander
here at Troy was Peteos' son, Menestheus.
No soldier born on earth could equal him
in battle at maneuvering men and horses --
though Nestor rivaled him, by grace of age.
In his command were Athens' fifty ships.

Great Aias led twelve ships from Salamis
and beached them where Athenians formed for battle.

Then there were those of Argos, those of Tiryns,
fortress with massive walls, Hermione
and Asine that lie upon the gulf,
Troizen, Eionai, the vineyard country
of Epidauros, Aigina and Mases:
these Diomedes, lord of the battle cry,
commanded with his comrade, Sthenelos,
whose father was illustrious Kapaneus,

and in third place Euryalos, a figure,
godlike in beauty, son of Mekisteus,
Lord Talaonides. Over them all
ruled Diomedes, lord of the battle cry,
and eighty black ships crossed the sea with these.
Next were the men who held the well-built city,
Mykenai, and rich Korinth, and Kleonai,
and Orneiai and fair Araithyrea
and Sikyon where first Adrestos ruled;
Hyperesia, hilltop Gonoessa,
Pellene, and the country round Aigion,
and those who held the north coast, Aigialos,
with spacious Helike. Their hundred ships
were under the command of Agamemnon,
Son of Atreus: he it was who led
by far the greatest number and the best,
And glorying in arms he now put on
a soldier's bronze -- distinguished amid heroes
for valor and the troops he led to war.

Next, those of Lakedaimon, land of gorges,
men who had lived at Pharis, Sparta, Messe
haunted by doves, Bryseiai, fair Augeiai,
Amyklai, too, and Helos by the sea,
and Laas and the land around Oitylos:
these Menelaos Agamemnon's brother,
lord of the war cry, led with sixty ships,
and drawn up separately from all the rest
they armed, as Menelaos on his own
burned to arouse his troops to fight. He burned
to avenge the struggles and the groans of Helen.

Next came the men of Pylos and Arene,
that trim town, and Thryon where they ford
Alpheios river, Aipyhighastoney,
and Pteleos and Helos -- Dorion, too,
where once the Muses, meeting Thamyris,
the Thracian, on his way from Oikhalia --
he could out sing the very Muses daughters.
Pride had made him say ill
of Zeus who bears the storm cloud for a shield.
For this affront they blinded him, bereft him
of his god-given song, and stilled his harping.
The countrymen of Pylos were commanded
by Nestor of Gerenia, charioteer.
Whose ninety decked ships lined the shore.

Then came
the troops who had their homes in Arkadia
under Kyllene crag: close-order fighters
who lived around the tomb of Aipytos,
at Pheneos, and at Orkhomenos
where there are many flocks; at Rhipe, too,
at Stratie, and in the windy town
Enispe; men of Tegee and lovely
Mantinea; men of Stymphalos
and Parrhasie, all of whom were led
by Agapenor, son of Angkhaios,
and he commanded sixty ships. Arkadains
able in war had thronged to go aboard,
for the Lord Marshal Agamemnon lent
those ships in which they crossed the wine dark sea,
as they had none, nor knowledge of seafaring.

Next were the soldiers from Bouprasion
and gracious Elis -- all that plain confined
by Hyrmine and Myrsinos, and by
Alesion and the Olenian Rock.
These had four captains with ten ships apiece,
on which the Epeioi had embarked in throngs.
Some served under Amphimakhos and Thalpios,
grandsons of Aktor, sons of Kteatos
and of Eurytos. Powerful Diores
Amaryngkeides commanded others,
and Polyxeinos led the fourth division --
son Agasthenes Augeiades.
Then came the islanders from Doulikhion
and the Ekhinades, all those who dwelt
opposite Elis, over the open sea,
Meges their captain, Meges Phyleides
begotten by that friend of Zeus, the horseman
Phyleus, who withdrew to Doulikhion
in anger at his father long ago.
Forty black ships had crossed the sea with Meges.

Odysseus, then, commanded the brave men
of Kephallenia: islanders of Ithaca
and Neritos whose leafy heights the sea wind
ruffles, and the men of Krokyleia,
of Aigilips the rocky isle, and those
of Samos and Zakynthos: those as well
who held the mainland eastward of the islands.
Odysseus, peer of Zeus in forethought, led them
in twelve good ships with cheek-paint at the bows.

Thoas, Andraimon's son, led the Aitolians,
inhabitants of Pleuron, Olenos,
Pylene, seaside Khalkis, Kalydon
of rocky mountainsides: Thoas their leader
 because the sons of Oineus were no more,
and red-haired Meleagros too was dead.
Command of all had thus devolved on Thoas,
and forty black ships crossed the sea with him.

Idomeneus, famed as a spear-fighter,
led the Cretans: all who came from Knossos,
Gotyn, the town of many walls, and Lyktos,
Miletos and Lykastos, gleaming white,
Phaistos and Rhytion, those pleasant towns --
all from that island of a hundred cities
served under Idomeneus, the great spearman,
whose second in command, Meriones,
fought like the slaughtering god of war himself.
Eighty black ships had crossed the sea with these.

Tlepolemos, the son of Herakles,
had led nine ships from Rhodes: impetuous men,
the Rhodians, in three regional divisions:
Lindos, Ielysos, and bright Kameiros,
serving under Tlepolemos, the spearman,
whose mother, Astyokheia, had been taken
by Herakles, who brought her from Ephyra
out of the Selleeis river valley,
where he had plundered many noble towns.
No sooner was Tlepolemos of age
than he had killed his father's uncle, old
Likymnios, Alkmene's warrior brother,
and fitting out his ships in haste, he sailed
over the deep sea, taking many with him
in flight from other descendants of Herakles.
Wandering, suffering bitter days at sea,
he came at last to Rhodes. The island, settled
in townships, one for each of three great clans,
was loved by Zeus, ruler of gods and men,
and wondrous riches he poured out upon them.

Nireus had led three well-found ships from Syme --
Nireus, Aglaia's child by Lord Kharopos --
Nireus, of all Danaans before Troy
most beautifully made, after Achilles,
a feeble man, though, with a small contingent.

Then those of Nisyros and Karpathos
and Kasos and the island town of Kos,
ruled by Eurypylos, and the Kalydnai,
island ruled by Pheidippos and Antiphos,
the sons of Thessalos, a son of Herakles.
Thirty long ships in line belonged to these.

Tell me now, Muse, of those from that great land
called Argos of Pelasgians, who lived
at Alos, at Alope, and at Trekhis,
and those of Phthra, those of Hellas, lands
of lovely women: all those troops they called
the Myrmidons and Hellenes and Achaians,
led by Achilles, in their fifty ships.
But these made no advances now to battle,
since he was not on hand to dress their lines.
No, the great runner, Prince Achilles, lay
amid the ships in desolate rage
for Briseis, his girl with her soft tresses --
the prize he captured, fighting all the way,
from Lyrnessos after he stormed that town
and stormed the walls of Thebe, overthrowing
the spearmen, Mynes and Epistrophos,
sons of Euenos Selepiades.
For her his heart burned, lying there,
but soon the hour would come when he would rouse.

Next were the men of Phylake, and those
who held Pyrasos, garden of Demeter,
Iton, maternal town of grazing flocks,
Antron beside the water, and the beds
of meadow grass at Pteleos: all these
were under Protesilaos' command
when that intrepid fighter lived--
but black earth held him under now, and grieving
at Phylake with lacerated cheeks
his bride was left, his house unfinished there.
Plunging ahead from his long ship to be
first man ashore at Troy of all Achaians,
he had been brought down by a Dardan spear.
By no means were his troops without a leader,
though sorely missing him: they had Podarkes,
another soldier son of Iphiklos
Phylakids, master of many flocks --
Podarkes, Protesilaos' blood brother,
a younger man, less noble. But the troops
were not at all in want of a commander,

though in their hearts they missed the braver one.
Forty black ships had sailed along with him.

Next were the soldiers who had lived at Pherai
by the great lake: at Glaphyrai and Boibe,
and in the well-kept city, Iaolkos.
Of their eleven ships Admetos' son
Eumelos had command -- the child conceived
under Admetos by that splendid queen,
Alkestis, Pelias' most beautiful daughter.

Next, those of Methone and Thaumakie,
of rugged Olizon and Meliboia.
These in their seven ships had been commanded
at first by Philoktetes, the great archer.
Fifty oarsmen in every ship, they came
as expert archers to the Trojan war.
But he, their captain, lay on Lemnos isle
in anguish, where the Achaians had marooned him,
bearing the black wound of a deadly snake.
He languished there, but soon, beside the ships,
the Argives would remember and call him back.
Meanwhile his men were not without a leader
though missing Philoktetes: Medon led them,
Oileus' bastard son, conceived
by Rhene under Oileus, raider of cities.

Next were the men of Trike and Ithome,
that rocky-terraced town, and Oikhalia,
'the city of Eurytos: over these
two sons of old Asklepios held command --
both skilled in healing: Podaleirios
and Makhaon. Thirty decked ships were theirs.

Next were the soldiers from Ormenios
and from the river source at Hypereia;
those of Asterion and those below
Titanos' high snow-whitened peaks. Eurypylos,
Euaimon's shining son, led all of these,
with forty black ships under his command.

Then those who held Argissa and Gyrtone,
Orthe, Elone, and the limestone city
Olooson, led by a dauntless man,
Polypoites, the son of Peirithoos,
whom Zeus, the undying, fathered. Polypoites
had been conceived by gentle Hippodameia

under Peirithoos, that day he whipped
the shaggy centaurs out of Pelion --
routed them, drove them to the Aithikes.
Polypoites as co-commander had
Leonteus, son of Koronos Kaineides.
Forty black ships had crossed the sea with these.

Gouneus commanded twenty-two from Kyphos.
The Enienes and the brave Peraiboi
served under him: all who had had their homes
around Dodona in the wintry north
and in the fertile vale of Titaressos.
Lovely that gliding river that runs on
into the Peneios with silver eddies
and rides it for a while as clear as oil --
a branch of Styx, on which great oaths are sworn.

The Magnetes were led by Prothoos,
Tenthredon's son: by Peneios they lived
and round Mount Pelion's shimmering leafy sides.
Forty black ships had come with Prothoos.

These were the lords and captains of the Danaans.
But tell me, Muse, of all the men and horses
who were the finest, under Agamemnon?
As for the battle horses, those were best
that came from Pheres' pastures, and Eumelos
drove those mares, as fleet as birds -- a team
perfectly matched in color and in age,
and level to a hair across the cruppers.
Apollo of the silver bow had breed them
in Pereie as fearsome steeds of war.
Of all the fighting men, most formidable
was Aias Telamonios -- that is
while great Achilles raged apart. Achilles
towered above them all; so did the stallions
that drew the son of Peleus in the war.
But now, amid the slim seagoing ships
he lay alone and raged at Agamemnon,
marshal of the army. And his people,
along the shore above the breaking waves,
with discus throw and javelin and archery
sported away the time. Meanwhile their teams
beside the chariots tore and champed at clover
and parsley from the marshes; the war-cars
shrouded in canvas rested in the shelters;
and, longing for their chief, beloved in war,

the Myrmidons, idly throughout the camp,
drifted and took no part in that day's fighting.

But now the marching host devoured the plain
as thought it were a prairie fire; the ground
beneath it rumbled, as when Zeus the lord
of lightning bolts, in anger at Typhoeus,
lashes the earth around Einarimos,
where his tremendous couch is said to be.
So thunderously groaned the earth
under the trampling of their coming on,
and they consumed like fire the open plain.

Iris arrived now, running on the wind,
as messenger from Zeus beyond the storm cloud,
bearing the grim news to the men of Troy.
They were assembled, at the gates of Priam,
young men and old, all gathered there, when she
came near and stood to speak to them: her voice
most like the voice of Priam's son Polites.
Forward observer for the Trojans, trusting
his prowess as a sprinter, he had held
his post mid-plain atop the burial mound
of the patriarch Aisyetes; waiting there
to see the Achaians leave their camp and ships.
In his guise, she who runs upon the wind,
Iris, now spoke to Priam:
        "Sir, old sir,
will you indulge inordinate talk as always,
just as in peacetime? Frontal war's upon us!
Many a time I've borne a hand in combat,
but never have I seen the enemy
in such array, committed, every man,
uncountable as leaves, or grains of sand,
advancing on the city through the plain!
Hektor, you are the one I call on: take
action as I direct you: the allies
that crowd the great town speak in many tongues
of many scattered countries. Every company
should get its orders from its own commander;
Let him conduct the muster and the sortie!"

Hektor punctiliously obeyed the goddess,
dismissed the assembly on her terms, and troops
ran for their arms. All city gates, wide open,
yawned, and the units poured out, foot soldiers,
horses and chariots, with tremendous din.

Rising in isolation on the plain
in face of Troy, there is a ridge, a bluff
open on all sides: Briar Hill they call it.
Men do, that is; the immortals know the place
to be the Amazon Myrine's tomb.
Anchored on this the Trojans and allies
formed for battle.

        Tall, with helmet flashing,
Hektor, great son of Priam, led the Trojans,
largest of those divisions and the best,
who drew up now and armed, and hefted spears.

The Dardans were commanded by Aineias,
whom ravishing Aphrodite had conceived
under Ankhises in the vales of Ida,
lying, immortal, in a man's embrace.
His co-commanders were Antenor's sons,
both battle-wise, Arkhelokhos and Akamas.

Then those from Zeleia, the lower slope
of Ida -- Trojans, men of means who drank
the waters of Aisepos dark and still --
they served under Lykaon's shining son,
Pandaros, whom Apollo taught the bow.

Adresteia's men, those of the hinterland
of Apaisos, Pityeia, the crag
of Treria -- all these Adrestos led
with Amphios of the linen cuirass, both
sons of Merops Perkosios, the seer
profoundest of all seers: he had refused
to let them take the path of war --
man-wasting war -- but they were heedless of him,
driven onward by dark powers of death.

Then, too, came those who lived around Perkote,
at Praktion, at Sestos and Abydos,
and old Arisbe: Asios their captain,
Asios Hyretakides -- who drove
geat sorrel horses from Selleeis river.

Hippothoos, led the tough Pelasgians
from Larisa's rich plow land -- Hippothoos
and the young soldier Pykais, both sons
of the Pelasgian Lethos Teutamides.

Then Thracians from beyond the strait, all those

whom Helle's rushing water bounded there,
Akamas led, and the veteran Peiroos.

Son of Troizenos Keades, Euphemos
led the Kikones from their distant shore;
and those more distant archers, Paiones,
Pyraikhmes led from Amydon,
from Axios bemirroring all the plain.
The Paphlagonians followed Pyaimenes,
shaggy, great-hearted, from the wild mule country
of the Enetoi -- men who held Kytoros
and Sesamos and had their famous homes
on the Parthenios riverbanks, at Kromma.
Aigialos, and loft Erythinoi.

Odios and Epistrophos were captains
of Halizones from Alybe, far
eastward, where the mines of silver are.

The Mysians Khromis led, with Ennomos,
reader of bird flight; signs in flurrying wings
would never save him from the last dark wave
when he went down before the battering hands
of the great runner, Achilles, in the river,
with other Trojans slain.

               The Phrygians
were under Phorkys and Askanios
from distant Askanie -- ready fighters.

The Lydians, then, Meiones, had for leader
Mesthles and Antiphos; these were the sons
born by Gygaie lake to Talaimenes.
They led men bred in vales under Mount Tmolos.
Nastes commanded Karians in their own tongue,
men of Miletos, Phthiron's leafy ridge,
Maiandors' rills and peaks of Mykale.
All these Amphimakhos and Nastes led,
Nomion's shining children. Wearing gold,
blithe as a girl, Nastes had gone to war,
but gold would not avail the fool
to save him from a bloody end. Achilles
Aiakides would down him in the river,
taking his golden ornaments for spoil.

Sarpedon led the Lykians, with Glaukos,
from Lykie afar, from whirling Xanthos.

Book Sixteen - A ship Fired, A Tide Turned

That was the way the fighting went
for one seagoing ship. Meanwhile Patroklos
approached Achilles his commander, streaming
warm tears--like a shaded mountain spring
that makes a rock ledge run with dusky water.
Achilles watched him come, and felt a pang for him.
Then the great prince and runner said:

       Patroklos,
why all the weeping? Like a small girl child
who runs beside her mother and cries and cries
to be taken up, and catches at her gown
and will not let her go, looking up in tears
until she has her wish: that's how you seem,
Patroklos, winking out your glimmering tears.
Have you something to tell the Myrmidons
or me? Some message you alone have heard
from Phthia? But they say that Aktor's son,
Menoitos, is living still, and Peleus,
the son of Aiakos, lives on
amid his Myrmidons, If one of these
were dead, we should be grieved.
       Or is this weeping
over the Argives, seeing how they perish
at the long ships by their own bloody fault!
Speak out now, don't conceal it, let us share it."

And groaning, Patroklos, you replied:

"Achilles, prince and greatest of Achaians,
be forbearing. They are badly hurt.
All who were the best fighters are now lying
among the ships with spear or arrow wounds.
Diomedes, Tydeus' rugged son, was shot;
Odysseus and Agamemnon, the great spearman,
have spear wounds; Eurypylos
took an arrow shot deep in his thigh.
Surgeons with medicines are attending them
to ease their wounds.
       But you are a hard case,
Achilles! God forbid this rage you nurse
should master me. You and your fearsome pride!
What good will come of it to anyone, later,

unless you keep disaster from the Argives?
Have you not pity?
Peleus, master of horse, was not your father,
Thetis was not your mother! Cold grey sea
and sea-cliffs bore you, making a mind so harsh.
If in your heart you fear some oracle,
some word of Zeus, told by your gentle mother,
then send me out at least, and send me quickly,
give me a company of Myrmidons,
and I may be a beacon to Danaans!
Lend me your gear to strap over my shoulders;
Trojans then may take me for yourself
and break off battle, giving our worn-out men
a chance to breathe. Respites are brief in war.
We fresh troops with one battle cry might easily
push their tired men back on the town,
away from ships and huts."

        So he petitioned,
witless as a child that what he begged for
was his own death, hard death and doom.
        Achilles
out of his deep anger made reply:

"Hard words, dear prince, There is no oracle
I know of that I must respect, no word
from Zeus reported by my gentle mother.
Only this bitterness eats at my heart
when one man would deprive and shame his equal,
taking back his prize by abuse of power.
The girl whom the Achaians chose for me
I won by my own spear. A town with walls
I stormed and sacked for her. Then Agamemnon
stole her back, out of my hands, as though
I were some Vagabond held cheap.

        All that
we can let pass as being over and done with;
I could not rage forever. And yet, by heaven, I swore
I would not rest from anger till the cries
and clangor of battle reached my very ships!
But you, now, you can strap my famous gear
on your own shoulders, and then take command
of Myrimidons on edge and ripe for combat,
now that like a dark storm cloud the Trojans
have poured round the first ships, and Argive troops
have almost no room for maneuver left,

with nothing to their rear but sea. The whole
townful of Trojans joins in, sure of winning,
because they cannot see my helmet's brow
a flash in range of them. They'd fill the gullies
with dead men soon, in flight up through the plain,
if Agamemnon were on good terms with me.
As things are, they've outflanked the camp. A mercy
for them that in the hands of Diomedes
 no great spear goes berserk, warding death
from the Danaans! Not yet have I heard
the voice of Agamemnon, either, shouting
out of his hateful skull. The shout of Hektor,
the killer, calling Trojans, makes a roar
like breaking surf, and with long answering cries
they hold the whole plain where they drove the Achaians.
Even so defend the ships, Patroklos.
Attack the enemy in force, or they
will set the ships ablaze with whirling fire
and rob Achaians of their dear return.
Now carry out the purpose I confide,
so that you'll win great honor for me, and glory
among Danaans; then they'll send me back
my lovely girl, with bright new gifts as well.
Once you expel the enemy from the ships,
rejoin me here. If Hera's lord,
the lord of thunder, grants you the day's honor,
covet no further combat far from me
with Trojan soldiers. That way you'd deny me
recompense of honor. You must not,
for joy of battle, joy of killing Trojans,
carry the fight to Ilion! Some power
out of Olympus, one of the immortal gods,
might intervene for them. The Lord Apollo
loves the Trojans. Turn back, then, as soon
as you restore the safety of the ships, a
and let the rest contend, out on the plain.
Ah, Father Zeus, Athena, and Apollo!
If not one Trojan of them all
should get away from death, and not one Argive
save ourselves were spared, we two alone
could pull down Troy's old coronet of towers!"

These were the speeches they exchanged. Now Aias
could no longer hold: he was dislodged
by spear-throws, beaten by the mind of Zeus
and Trojan shouts. His shining helm rang out
around his temples dangerously with hits

as his helm plates were struck and struck again;
he felt his shoulder galled on the left side
hugging the glittering shield -- and yet they could not
shake it, putting all their weight in throws.
In painful gasps his breath came, sweat ran down
in rivers off his body everywhere;
no rest for him, but trouble upon trouble.

Now tell me, Muses, dwellers on Olympus,
how fires first fell on the Achaian ships!
Hektor moved in to slash with his long blade
at Aias' ash wood shaft, and near the spearhead
lopped it off. Then Telamonian Aias
wielded a pointless shaft, while far away
the flying bronze head rang upon the ground,
and Aias shivered knowing in his heart
the work of gods: how Zeus, the lord of thunder,
cut off his war-craft in that fight, and willed
victory to the Trojans. He gave way
before their missiles as they rushed in throwing
untiring fire into the ship. It caught
at once, a-gush with flame, and fire lapped
about the stern.
                Achilles smote his thighs
and said to Patroklos:
                "Now go into action,
prince and horseman! I see roaring fire
burst at the ships. Action, or they'll destroy them,
leaving no means of getting home. Be quick,
strap on my gear, while I alert the troops!"

Patroklos now put on the flashing bronze.
Greaves were the first thing, beautifully fitted
to calf and shin with silver ankle chains;
and next he buckled round his ribs the cuirass,
blazoned with stars, of swift Aiakides;
then slung the silver-studded blade of bronze
about his shoulders, and the vast solid shield;
then on his noble head he placed the helm,
its plume of terror nodding high above,
and took two burly spears with his own handgrip.
He did not take the great spear of Achilles,
weighty, long, and tough. No other Achaian
had the strength to wield it, only Achilles.
It was a Pelian ash, cut on the crest
of Pelion, given to Achilles' father
by Kheiron to deal death to soldiery.

He then ordered his war-team put in harness
by Automedon, whom he most admired
after Prince Achilles, breaker of men,
for waiting steadfast at his call in battle.
Automedon yoked the fast horses for him --
Xanthos and Balios, racers of wind.
The storm gust Podarge, who once had grazed
green meadowland by the Ocean stream, conceived
and bore them to the west wind, Zephyros.
In the side-traces Pedasos, a thoroughbred,
was added to the team; Achilles took him
when he destroyed the city of Eetion.
Mortal, he ran beside immortal horses.
Achilles put the Myrmidons in arms,
the whole detachment near the huts. Like wolves,
carnivorous and fierce and tireless,
who rend a great stag on a mountain side
and feed on him, their jaws reddened with blood,
loping in a pack to drink spring water,
lapping the dark rim up with slender tongues,
their chops a-drip with fresh blood, their hearts
unshaken ever, and their bellies glutted:
such were the Myrmidons and their officers,
running to form up round Achilles' brave
companion-in-arms.
        And like the god of war
among them was Achilles: he stood tall
and sped the chariots and shield men onward.

Fifty ships there were that Lord Achilles,
favored of heaven, led to Troy. In each
were fifty soldiers, shipmates at the rowlocks.
Five he entrusted with command and made
lieutenants, while he ruled them all as king.
One company was headed by Menesthios
in his glittering breastplate, son of Sperkheios,
a river fed by heaven. Peleus' daughter,
beautiful Polydore, had conceived him
lying with Sperkheios, untiring stream,
a woman with a god; but the world thought
she bore her child to Perieres' son,
Boros, who married her in the eyes of men
and offered countless bridal gifts. A second
company was commanded by Eudoros,
whose mother was unmarried: Polymele,
Phylas; daughter, a beautiful dancer
with whom the strong god Hermes fell in love,

seeing her among singing girls who moved
in measure for the lady of belling hounds,
Artemis of the golden shaft. And Hermes,
pure Deliverer, ascending soon
to an upper room, lay secretly with her
who was to bear his brilliant son, Eudoros,
a first-rate man at running and in war.
When Eileithyia, sending pangs of labor,
brought him forth to see the sun rays, then
strong-minded Ekhekleos, Aktor's son,
led the girl home with countless bridal gifts;
but Phylas in his age brought up the boy
with all kind care, as though he were a son.
Company three was led by Peisandros
Maimalides, the best man with a spear,
of all Myrmidons after Patroklos.
Company four the old man, master of horse,
Phoinix, commanded. Alikimedon, son
of Laerkes, commanded company five.
When all were mustered under their officers,
Achilles had strict orders to impart:

"Myrmidons, let not one man forget
how menacing you were against the Trojans
during my anger and seclusion: how
each one reproached me, saying, 'Iron hearted
son of Peleus, now we see: your mother
brought you up on rage, merciless man,
the way you keep your men confined to camp
against their will! We might as well sail home
in our seagoing ships, now this infernal
anger has come over you!' That way
you often talked, in groups around our fires.
Now the great task of battle is at hand
that you were longing for! Now every soldier
keep a fighting heart and face the Trojans!"

He stirred and braced their spirit; every rank
fell in more sharply when it heard its king.
As when a builder fitting stone on stone
lays well a high house wall to buffet back
the might of winds, just so
they fitted helms and studded shields together:
shield-rim on shield-rim, helmet on helmet, men
all pressed on one another, horse hair plumes
brushed on the bright crests as the soldiers nodded,
densely packed as they were.

Before them all
two captains stood in gear of war: Patroklos
and Automedon, of one mind, resolved
to open combat in the lead.
Achilles
went to his hut. He lifted up the lid
of a sea chest, all intricately wrought,
that Thetis of the silver feet had stowed
aboard his ship for him to take to Ilion,
filled to the brim with shirts, wind-breaking cloaks,
and fleecy rugs. His hammered cup was there,
from which no other man drank the bright wine,
and he made offering to no god but Zeus.
Lifting it from the chest, he purified it
first with brimstone, washed it with clear water,
and washed his hands, then dipped it full of wine.
Now standing in the forecourt, looking up
toward heaven, he prayed and poured his offering out,
and Zeus who plays in thunder heard his prayer:

"Zeus of Dodona, god of Pelasgains,
O god whose home lies far! Ruler of wintry
harsh Dodona! Your interpreters,
the Selloi, live with feet like roots, unwashed,
and sleep on the hard ground. My lord, you heard me
praying before this, and honored me
by punishing the Achaian army. Now,
again, accomplish what I most desire.
I shall stay on the beach, behind the ships,
but send my dear friend with a mass of soldiers,
Myrmidons, into combat. Let your glory,
Zeus who views the wide world, go beside him.
Sir, exalt his heart,
so Hektor too may see whether my friend
can only fight when I am in the field,
or whether single handed he can scatter them
before his fury! When he has thrown back
their shouting onslaught from the ships, then let him
return unhurt to the shipways and to me
his gear intact with all his fighting men."

That was his prayer, and Zeus who views the wide world
heard him. Part he granted, part denied:
he let Patroklos push the heavy fighting
back from the ships, but would not let him come
unscathed from battle.

Now, after Achilles
had made his prayer and offering to Zeus,
he entered his hut again, restored the cup
to his sea chest, and took his place outside--
desiring still to watch the savage combat
of Trojans and Achaians. Brave Patroklos'
men moved forward with high hearts until
they charged the Trojans -- Myrmidons in waves,
like hornets that small boys, as boys will do,
the idiots, poke up with constant teasing
in their daub chambers on the road,
to give everyone trouble. If some traveler
who passes unaware should then excite them,
all the swarm comes raging out
to defend their young. So hot, so angrily
the Myrmidons came pouring from the ships
in a quenchless din of shouting. And Patroklos
cried above them all:

          "O Myrmidons,
brothers-in-arms of Peleus' son, Achilles,
fight like men, dear friends, remember courage,
let us win honor for the son of Peleus!
He is the greatest captain on the beach,
his officers and soldiers are the bravest!
Let King Agamemnon learn his folly
in holding cheap the best of the Achaians!"

Shouting so, he stirred their hearts. They fell
as one man on the Trojans, and the ships
around them echoed the onrush and the cries.
On seeing Menoitios' powerful son, and with him
Automedon, a flash with brazen gear,
the Trojan ranks broke, and they caught their breath,
imagining that Achilles the swift fighter
had put aside his wrath for friendship's sake.
Now each man kept an eye out for retreat
from sudden death. Patroklos drove ahead
against their center with his shining spear,
into the huddling mass, around the stern
of Protesilaos' burning ship. He hit
Pyraikhmes, who had led the Paiones
from Amydon, from Axios' wide river --
hit him in the right shoulder. Backward in dust
he tumbled groaning, and his men-at-arms,
the Paiones, fell back around him. Dealing
death to a chief and champion, Patroklos

drove them in confusion from the ship,
and doused the tigerish fire. The hull half-burnt
lay smoking on the shipway. Now the Trojans
with a great outcry streamed away; Danaans
poured along the curved ships, and the din
of war kept on. As when the lightning master,
Zeus, removes a dense cloud from the peak
of some great mountain, and the lookout points
and spurs and clearings are distinctly seen
as though pure space had broken through from heaven:
so when the dangerous fire had been repelled
Danaans took breath for a space. The battle
had not ended, though; not yet were Trojans
put to rout by the Achaian charge
or out of range of the black ships. They withdrew
but by regrouping tried to make a stand.
                    In broken
ranks the captains sought and killed each other,
Menoitios' son making the first kill.
As Areilykos wheeled around to fight,
he caught him with his spearhead in the hip,
and drove the bronze through, shattering the bone.
He sprawled face downward on the ground.
                    Now veteran
Menelaos thrusting past the shield
of Thoas to the bare chest brought him down.
Rushed by Amphiklos, the alert Meges
got his thrust in first, hitting his thigh
where a man's muscles bunch. Around the spear head
tendons were split, and darkness veiled his eyes.
Nestor's sons were in action: Antilokhos
with his good spear brought down Atymnios,
laying open his flank; he fell head first.
Now Maris moved in, raging for his brother,
lunging over the dead man with his spear,
but Thrasymedes had already lunged
and did not miss, but smashed his shoulder squarely,
tearing his upper arm out of the socket,
severing muscles, breaking through the bone.
He thudded down and darkness veiled his eyes.
So these two, overcome by the two brothers,
dropped to the underworld of Erebos.
They were Sarpedon's true brothers-in-arms
and sons of Amisodaros, who reared
the fierce Khimaira, nightmare to may men.
Alas, Oileus' son, drove at Kleoboulos
and took him alive, encumbered in the press,

but killed him on the spot with a sword stroke
across his nape -- the whole blade running hot
with blood, as welling death and his harsh destiny
possessed him. Now Peneleos
and Lykon clashed; as both had cast and missed
and lunged and missed with spears,
they fought again with swords. The stroke of Lykon
came down on the other's helmet ridge
but his blade broke at the hilt. Peneleos
thrust at his neck below the ear and drove
the blade clear in and through; his head toppled,
held only by skin, and his knees gave way.
Meriones on the run overtook Akamas
mounting behind his horses and hit his shoulder,
knocking him from the car. Mist swathed his eyes,
Idomeneus thrust hard at Erymas' mouth
with his hard bronze. The spearhead passed on through
beneath his brain and split the white brain-pan.
His teeth were dashed out, blood filled both his eyes,
and from his mouth and nostrils as he gaped
he spurted blood. Death's cloud enveloped him.
There each Danaan captain killed his man.
As ravenous wolves come down on lambs and kids
astray from some flock that in hilly country
splits in two by a shepherd's negligence,
and quickly wolves bear off the defenseless things,
so when Danaans fell on Trojans, shrieking
flight was all they thought of, not of combat.
Aias the Tall kept after bronze-helmed Hektor,
casting his lance, but Hektor, skilled in war,
would fit his shoulders under the bull's-hide shield,
and watch for whizzing arrows, thudding spears.
Aye, though he knew the tide of battle turned,
he kept his discipline and saved his friends.
As when Lord Zeus would hang the sky with storm,
a cloud may enter heaven from Olympus
out of crystalline space, so terror and cries
increased about the shipways. In disorder
men withdrew. Then Hektor's chariot team
cantering bore him off with all his gear,
leaving the Trojans whom the moat confined;
and many chariot horses in that ditch,
breaking their poles off at the tip, abandoned
war-cars and masters. Hard on their heels
Patroklos kept on calling all Danaans
onward with slaughter in his heart. The Trojans,
yelling and clattering, filled all the ways,

their companies cut in pieces, High in air
a blast of wind swept on, under the clouds,
as chariot horses raced back toward the town
away from the encampment. And Patroklos
rode shouting where he saw the enemy mass
in uproar: men fell from their chariots
under the wheels and cars jounced over them,
and running horses leapt over the ditch--
immortal horses, whom the gods gave Peleus,
galloping as their mettle called them onward
after Hektor, target of Patroklos.
But Hektor's battle-team bore him away.

As under a great storm black earth is drenched
on an autumn day, when Zeus pours down the rain
in scudding gusts to punish men, annoyed
because they will enforce their crooked judgments
and banish justice from the market place,
thoughtless of the gods' vengeance; all their streams
run high and full, and torrents cut their way
down dry declivities into the swollen sea
with a hoarse clamor, headlong out of hills,
while cultivated fields erode away --
such was the gasping flight of the Trojan horses.

When he had cut their first wave off, Patroklos
forced it back again upon the ships
as the men fought toward the city. In between
the ships and river and the parapet
he swept among them killing, taking toll
for many dead Achaians.   First,
thrusting past Pronoos' shield, he hit him
on the bare chest, and made him crumple: down
he tumbled with a crash. Then he rushed Thestor,
Enop's son, who sat all double up
in a polished war-car, shocked out of his wits,
the reins flown from his hands -- and the Achaian
got home his thrust on the right jawbone, driving
through his teeth. He hooked him by the spearhead
over the chariot rail, as a fisherman
on a point of rock will hook a splendid fish
with line and dazzling bronze out of the ocean:
so from his chariot on the shining spear
he hooked him gaping and face downward threw him
life going out of him as he fell.
   Patroklos
now met Erylaos' rush and hit him square

mid-skull with a big stone. Within his helm
the skull was cleft asunder, and down he went
headfirst to earth; heartbreaking death engulfed him.
Next Erymas, Amphoteros, Epaltes,
Tlepolemos Damastorides, Ekhios,
Pyris, Ipheus, Euippos, Polymelos,
all in quick succession he brought down
to the once peaceful pastureland.
      Sarpedon,
seeing his brother-in-arms in their unbelted
battle jackets downed at Patroklos' hands,
called in bitterness to the Lykians:

"Shame, O Lykians, where are you running?
Now you show your speed!
     I'll take on this one,
and learn what man he is that has the power
to do such havoc as he has done among us,
cutting down so many, and such good men."

He vaulted from his car with all his gear,
and on his side Patroklos, when he saw him,
leapt from his car. Like two great birds of prey
with hooked talons and angled beaks, who screech
and clash on a high ridge of rock, these two
rushed one another with hoarse cries. But Zeus,
the son of crooked-minded Kronos, watched,
and pitied them. He said to Hera:
     "Ai!"
Sorrow for me, that in the scheme of things
the dearest of men to me must lie in dust
before the son of Menoitios, Patroklos.
My heart goes two ways as I ponder this:
shall I catch up Sarpedon
out of the mortal fight with all its woe
and put him down alive in Lykia,
in that rich land? Or shall I make him fall
beneath Patroklos' hard-thrown spear?"
     Then Hera
of the wide eyes answered him:

     "O fearsome power,
my Lord Zeus, what a curious thing to say.
A man who is born to die, long destined for it,
would you set free from that unspeakable end?
Do so; but not all of us will praise you.
And this, too, I may tell you: ponder this:

should you dispatch Sarpedon home alive,
anticipate some other god's desire
to pluck a man he loves out of the battle,
Many who fight around the town of Priam
sprang from immortals; you'll infuriate these.
No, dear to you though he is, and though you mourn him,
let him fall, even so, in the rough battle,
killed by the son of Menoitios, Patroklos.
Afterward, when his soul is gone, his lifetime
ended, Death and sweetest Sleep can bear him
homeward to the broad domain of Lykia.
There friends and kin may give him funeral
with tomb and stone, the trophies of the dead."

To this the father of gods and men agreed,
but showered bloody drops upon the earth
for the dear son Patroklos would destroy
in fertile Ilion, far from his home.
When the two men had come in range, Patroklos
turned like lightning against Thrasydemos,
a tough man ever at Sarpedon's side,
and gave him a death-wound in the underbelly.
Sarpendon's counterthrust went wide, but hit
the trace horse, Pedasos, in the right shoulder.
Screaming harshly, panting his life away,
he crashed and whinnied in the dust; the spirit
left him with a wingbeat. The team shied
and strained apart with a great creak of the yoke
as reins were tangled over the dead weight
of their outrider fallen. Automedon,
the good soldier, found a way to end it:
pulling his long blade from his hip
he jumped in fast and cut the trace horse free.
The team then arranged themselves beside the pole,
drawing the reins taut, and once more,
devoured by fighting madness, the two men clashed.
Sarpedon missed again. He drove his spearhead
over the left shoulder of Patroklos,
not even grazing him. Patroklos then
made his last throw, and the weapon left his hand
with flawless aim. He hit his enemy
just where the muscles of the diaphragm
encased his throbbing heart. Sarpedon fell
the way an oak or poplar or tall pine
goes down, when shipwrights in the wooded hills
with whetted axes chop it down for timber.
So, full length, before his war-car lay

Sarpedon raging, clutching the bloody dust.
Imagine a greathearted sultry bull
a lion kills amid a shambling herd:
with choking groans he dies under the claws.
So, mortally wounded by Patroklos
the chief of Lykian shields men lay in agony
and called his friend by name:

       "Glaukos, old man,
old war-dog, now's the time to be a spearman!
Put your heart in combat! Let grim war
be all your longing! Quickly, if you can,
arouse the Lykian captains, round them up
to fight over Sarpedon. You, too, fight
to keep my body, else in later days
this day will be your shame. You'll hang your head
all your life long, if these Achaians take
my armor here, where I have gone down fighting
before the ships. Hold hard; cheer on the troops!"

The end of life came on him as he spoke,
closing his eyes and nostrils. And Patroklos
with one foot on his chest drew from his belly
spearhead and spear; the diaphragm came out,
so he extracted life and blade together.
Myrmidons clung to the panting Lykian horses,
rearing to turn the car left by their lords.
But bitter anguish at Sarpedon's voice
had come to Glaukos, and his heart despaired
because he had not helped his friend. He gripped
his own right arm and squeezed it, being numb
where Teukros with a bowshot from the rampart
had hit him while he fought for his own men,
and he spoke out in prayer to Lord Apollo:

"Hear me, O lord, somewhere in Lykian farmland
or else in Troy: for you have power to listen
the whole world round to a man hard pressed as I!
I have my sore wound, all my length of arm
a-throb with lancing pain; the flow of blood
cannot be stanched; my shoulder's heavy with it.
I cannot hold my spear right or do battle,
cannot attack them. Here's a great man destroyed,
Sarpedon, son of Zeus. Zeus let his own son
die undefended. O my lord, heal this wound,
lull me my pains, put vigor in me! Let me
shout to my Lykians, move then into combat!

Let me give battle for the dead man here!"

This way he prayed, and Phoibos Apollo heard him,
cutting his pain and making the dark blood dry
on his deep wound, then filled his heart with valor.
Glaukos felt the change, and knew with joy
how swiftly the great god had heard his prayer.
First he appealed to the Lykian captains, going
right and left, to defend Sarpedon's body
then on the run he followed other Trojans,
Poulydamas, Panthoos' son, Agenor,
and caught up with Aineias and with Hektor,
shoulder to shoulder, urgently appealing:

"Hektor, you've put your allies out of mind,
those men who give their lives here for your sake
so distant from their friends and lands: you will not
come to their aid! Sarpedon lies there dead,
commander of the Lykians, who kept
his country safe by his firm hand, in justice!
Ares in bronze has brought him down: the spear
belonged to Patroklos. Come, stand with me, friends,
and count it shame if they strip off his gear
or bring dishonor on his body -- these
fresh Myrmidons enraged for the Danaans
cut down at the shipways by our spears!"

At this, grief and remorse possessed the Trojans,
grief not to be borne, because Sarpedon
had been a bastion of the town of Troy,
foreigner though he was. A host came with him,
but he had fought most gallantly of all.
They made straight for the Danaans, and Hektor
led them, hot with anger for Sarpedon.
Patroklos in his savagery cheered on
the Achaians, first the two named Aias, both
already aflame for war:
          "Aias and Aias,
let it be sweet to you to stand and fight!
You always do; be lionhearted, now.
The man who crossed the rampart of Achaians
first of all lies dead: Sarpedon. May we
take him, dishonor him, and strip his arms,
and hurl any friend who would defend him
into the dust with our hard bronze!"

At this they burned to throw the Trojans back.

And both sides reinforced their battle lines,
Trojans and Lykians, Myrmidons and Achaians,
moving up to fight around the dead
with fierce cries and clanging of men's armor.

Zeus unfurled a deathly gloom of night
over the combat, making battle toil
about his dear son's body a fearsome thing.
At first, the Trojans drove back the Achaians,
fiery-eyed as they were; one Myrmidon,
and not the least, was killed: noble Epeigeus,
a son of Agakles. In Boudeion,
a flourishing town, he ruled before the war,
but slew a kinsman. So he came as suppliant
to Peleus and to Thetis, who enlisted him
along with Lord Achilles, breaker of men,
To make war in the wild-horse country of Ilion
against the Trojans. Even as he touched the dead man,
Hektor hit him square upon the crest
with at great stone: his skull split in the helmet,
and he fell prone upon the corpse. Death's cloud
poured round him, heart-corroding. Grief and pain
for this friend dying came to Lord Patroklos,
who pounced through spear-play like a diving hawk
that puts jackdaws and starlings wildly to flight:
straight through Lykians, through Trojans, too
you drove, Patroklos, master of horse,
in fury for your friend. Sthenelaos
the son of Ithaimenes was the victim:
Patroklos with a great stone broke his nape-cord.

Backward the line bent, Hektor too gave way,
as far as a hunting spear may hurtle, thrown
by a man in practice or in competition
or matched with deadly foes in war. So far
the Trojans ebbed, as the Achaians drove them.
Glaukos, commander of Lykians, turned first,
to bring down valorous Bathykles, the son
of Khalkon, one who had his home in Hellas,
fortunate and rich among the Myrmidons.
Whirling as this man caught him, Glaukos hit him
full in the breastbone with his spear, and down
he thudded on his face. The Achaians grieved
to see their champion fallen, but great joy
came to the Trojans, and they thronged about him.
Not that Achaians now forgot their courage,
no, for their momentum carried them on.

Meriones brought down a Trojan soldier,
Laogonos, Onetor's rugged son,
a priest of Zeus on Ida, honored there
as gods are. Gashed now under jaw and ear
his life ran out, and hateful darkness took him.
Then at Meriones Aineias cast
his bronze-shod spear, thinking to reach his body
under the shield as he came on. But he
looked out for it and swerved, slipping the spear-throw,
bowing forward, so the long shaft stuck
in earth behind him and the butt quivered;
the god Ares deprived it of its power.
Aineias raged and sneered:

     "Meriones,
fast dodger that you are, if I had hit you
my spearhead would have stopped your dance for good!"
Meriones, good spearman, answered him:

"For all your power, Aineias, you could hardly
quench the fighting spirit of every man
defending himself against you. You are made
of mortal stuff like me. I, too, can say,
if I could hit you square, then tough and sure
as you may be, you would concede the game
and give your soul to the lord of nightmare, Death."

Patroklos said to him sharply:

     "Meriones,
you have your skill, why make a speech about it?
No, old friend, rough words will make no Trojans
back away from the body. Many a one
will be embraced by earth before they do.
War is the use of arms, words are for council.
More talk's pointless now; we need more fighting!"

He pushed on, and godlike Meriones
fought at his side. Think of the sound of strokes
woodcutters make in mountain glens, the echoes
ringing for listeners far away: just so
the battering din of these in combat rose
from earth where the living go their ways -- the clang
of bronze, hard blows on leather, on bull's hide,
as long sword blades and spearheads met their marks.
And an observer could not by now have seen
the Prince Sarpedon, since from head to foot

he lay enwrapped in weapons, dust, and blood.
Men kept crowding around the corpse. Like flies
that swarm and drone in farmyards round the milk pails
on spring days, when the pails are splashed with milk:
just so they thronged around the corpse. And Zeus
would never turn his shining eyes away
from this melee, but watched then all and pondered
long over the slaughter of Patroklos --
whether in that place, on Sarpedon's body,
Hektor should kill the man and take his gear,
or whether he, Zeus, should augment the moil
of battle for still other men. He weighted it
and thought this best: that for a while Achilles
shining brother-in-arms should drive his foes
and Hektor in the bronze helm toward the city,
taking the lives of many. First of all
he weakened Hektor, made him mount his car
and turn away, retreating, crying out
to others to retreat: for he perceived
the dipping scales of Zeus. At this the Lykians
themselves could not stand fast, but all turned back,
once they had seen their king struck to the heart,
lying amid swales of dead -- for many
fell to earth beside him when Lord Zeus
had drawn the savage battle line. So now
Achaians lifted from Sarpedon's shoulders
gleaming arms of bronze, and these Patroklos
gave to his soldiers to be carried back
to the decked ships. At this point, to Apollo
Zeus who gathers cloud said:

wipe away the blood mantling Sarpedon;
take him up, out of the play of spears,
a long way off, and wash him in the river,
anoint him with ambrosia, put ambrosial
clothing on him. Then have him conveyed
by those escorting spirits quick as wind,
sweet Sleep and Death, who are twin brothers. These
will set him down in the rich broad land of Lykia,
and there his kin and friends may bury him
with tomb and stone, the trophies of the dead."

Attentive to his father, Lord Apollo
went down the foothills of Ida to the field
and lifted Prince Sarpedon clear of it,
He bore him far and bathed him in the river,
scented him with ambrosia, put ambrosial

clothing on him, then had him conveyed
by those escorting spirits quick as wind,
sweet Sleep and Death, who are twin brothers. These
returned him to the rich broad land of Lykia.

Patroklos, calling to his team, commanding
Automedon, rode on after the Trojans
and Lykains -- all this to his undoing,
the blunderer. By keeping Achilles' mandate,
he might have fled black fate and cruel death.
But overpowering is the mind of Zeus
forever, matched with man's. He turns in fright
the powerful man and robs him of his victory
easily, though he drove him on himself.
So now he stirred Patroklos' heart to fury.

Whom first, whom later did you kill in battle,
Patroklos, when the gods were calling deathward?
First it was Adrestos, Autonoos,
and Ekheklos; then Perimos, Megades,
Eristor, Melanippos; afterward,
Elasos, Moulios, Pylartes. These
he cut down, while the rest looked to their flight.
Troy of the towering gates was on the verge
of being taken by the Achaians, under
Patroklos; drive; he raced with blooded spear
ahead and around it. On the massive tower
Phoibos Apollo stood as Troy's defender,
deadly toward him. Now three times Patroklos
assaulted the high wall at the tower joint,
and three times Lord Apollo threw him back
with counterblows of his immortal hands
against the resplendent shield. The Achaian then
a fourth time flung himself against the wall,
more than human in fury. But Apollo
thundered:
          "Back, Patroklos, lordly man!

Destiny will not let this fortress town
of Trojans fall to you! Not to Achilles,
either, greater far though he is in war!"
Patroklos now retired, a long way off
and out of range of Lord Apollo's anger.
Hektor had held his team at the Skaian Gates,
being of two minds: should he re-engage,
or call his troops to shelter behind the wall?
While he debated this, Phoibos Apollo

stood at his shoulder in a strong man's guise:
Asios, his maternal uncle, brother
of Hekabe and son of Dymas, dweller
in Phrygia on Sangarios river.
Taking his semblance now, Apollo said:

"Why break off battle, Hektor? You need not.
Were I superior to you in the measure
that I am now inferior, you'd suffer
from turning back so wretchedly from battle.
Action! Lash your team against Patroklos,
and see if you can take him. May Apollo
grant you the glory!"

       And at this, once more
he joined the melee, entering it as a god.
Hektor in splendor called Kebriones
to whip the horses toward the fight. Apollo,
disappearing in to the ranks, aroused
confusion in the Argives, but on Hektor
and on the Trojans he conferred his glory.
Letting the rest go, Hektor drove his team
straight at Patroklos; and Patroklos faced him
vaulting from his war-car, with his spear
gripped in his left hand; in his right
he held enfolded a sparkling jagged stone.
Not for long in awe of the other man,
he aimed and braced himself and threw the stone
and scored a direct hit on Hektor's driver,
Kebriones, a bastard son of Priam,
smashing his forehead with the jagged stone.
Both brows were hit at once, the frontal bone
gave way, and both his eyes burst from their sockets
dropping into the dust before his feet,
as like a diver from the handsome car
he plummeted, and life ebbed from his bones.
You jeered at him then, master of horse, Patroklos:

"God, what a nimble fellow, somersaulting!
If he were out at sea in the fishing grounds
this man could feed a crew, diving for oysters,
going overboard even in rough water,
the way he took that earth-dive from his car.
The Trojans have their acrobats, I see."

With this, he went for the dead man with a spring
like a lion, one that has taken a chest wound

while ravaging a cattle pen -- his valor
his undoing. So you sprang, Patroklos,
on Kebriones. Then Hektor, too, leapt down
out of his chariot, and the two men fought
over the body like two mountain lions
over the carcass of a buck, both famished,
both in pride of combat. So these two
fought now for Kebriones, two champions,
Patroklos, son of Menoitios, and Hektor,
hurling their bronze to tear each other's flesh.
Hektor caught hold of the dead man's head and held,
while his antagonist clung to a single foot,
as Trojans and Danaans pressed the fight.
As south wind and the southeast wind, contending
in mountain groves, make all the forest thrash,
beech trees and ash trees and the slender cornel
swaying their pointed boughs toward one another
in roaring wind, and snapping branches crack:
so Trojans and Achaians made a din
as lunging they destroyed each other. Neither
considered ruinous flight. Many sharp spears
and arrows trued by feathers from the strings
were fixed in flesh around Kebriones,
and boulders crashed on shields, as they fought on
around him. And a dust cloud wrought
by a whirlwind hid the greatness of him slain,
minding no more the mastery of horses.
Until the sun stood at high noon in heaven,
spears bit on both sides, and the soldiers fell;
but when the sun passed toward unyoking time,
the Achaians outfought destiny to prevail.
Now they dragged off gallant Kebriones
out of range, away from the shouting Trojans,
to strip his shoulders of his gear. And fierce
Patroklos hurled himself upon the Trojans,
in onslaughts fast as Ares, three times, wild
yells in his throat. Each time he killed nine men.
But on the fourth demonic foray, then
the end of life loomed up for you, Patroklos.
Into the combat dangerous Phoibos came
against him, but Patroklos could not see
the god, enwrapped in cloud as he came near.
He stood behind and struck with open hand
the man's back and broad shoulders, and the eyes
of the fighting man were dizzied by the blow.
Then Phoibos sent the captain's helmet rolling
under the horses' hooves, making the ridge

ring out, and dirtying all the horse hair plume
with blood and dust. Never in time before
had this plumed helmet been befouled with dust,
the helmet that had kept a hero's brow
unmarred, shielding Achilles' head. Now Zeus
bestowed it upon Hektor, let him wear it,
though his destruction waited. For Patroklos
felt his great spear shaft shattered in his hands,
long, tough, well-shod, and seasoned though it was;
his shield and strap fell to the ground the Lord
Apollo, son of Zeus, broke off his cuirass.
Shock ran through him, and his good legs failed,
so that he stood agape. Then from behind
at close quarters, between the shoulder blades,
a Dardaan fighter speared him: Panthoos' son,
Euphorbos, the best Trojan of his age
at handling spears, in horsemanship and running:
he had brought twenty chariot fighters down
since entering combat in his chariot,
already skilled in the craft of war. This man
was first to wound you with a spear, Patroklos,
but did not bring you down. Instead, he ran back
into the melee, pulling from the flesh
his ashen spear, and would not face his enemy,
even disarmed, in battle. Then Patroklos,
disabled by the god's blow and the spear wound
moved back to save himself amid his men.
But Hektor, seeing that his brave adversary
tried to retire, hurt by the spear wound, charged
straight at him through the ranks and lunged for him
low in the flank, driving the spearhead through.
He crashed, and all Achaian troops turned pale.
Think how a lion in his pride brings down
a tireless boar; magnificently they fight
on a mountain crest for a small gushing spring --
both in desire to drink -- and by sheer power
the lion conquers the great panting boar:
that was the way the son of Priam, Hektor,
closed with Patroklos, son of Menoitios,
killer of many, and took his life away.
Then glorying above him he addressed him:

"Easy to guess, Patroklos, how you swore
to ravage Troy, to take the sweet daylight
of liberty from our women, and to drag them
off into ships to your own land -- you fool!
Between you and those women there is Hektor's

war-team, thundering out to fight! My spear
has pride of place among the Trojan warriors,
keeping their evil hour at bay.
The kites will feed on you, here on this field.
Poor devil, what has that great prince, Achilles,
done for you? He must have told you often
as you were leaving and he stayed behind,
'Never come back to me, to the deep sea ships,
Patroklos, till you cut to rags
the bloody tunic on the chest of Hektor!'
That must have been the way he talked, and won
Your mind to mindlessness."

In a low faint voice,
Patroklos, master of horse, you answered him:

"This is your hour to glory over me,
Hektor, The Lord Zeus and Apollo gave you
the upper hand and put me down with ease.
They stripped me of my arms. No one else did.
Say twenty men like you had come against me,
all would have died before my spear.
No, Leto's son and fatal destiny
have killed me; if we speak of men, Euphorbos.
You were in third place, only in at the death.
I'll tell you one thing more; take it to heart.
No long life is ahead for you. This day
your death stands near, and your immutable end,
at Prince Achilles' hands."
His own death
came on him as he spoke, and soul from body,
bemoaning severance from youth and manhood,
slipped to be wafted to the underworld.
Even in death Prince Hektor still addressed him:

"Why prophesy my sudden death, Patroklos?
Who knows, Achilles, son of bright-haired Thetis,
might be hit first; he might be killed by me."

At this he pulled his spearhead from the wound
setting his heel upon him; then he pushed him
over on his back, clear of the spear,
and lifting it at once sought Automedon,
companion of the great runner, Achilles,
longing to strike him. But the immortal horses,
gift of the gods to Peleus, bore him away.

Book 24 - A Grace Given in Sorrow

The funeral games were over. Men dispersed
and turned their thoughts to supper in their quarters,
then to the boon of slumber. But Achilles
thought of his friend, and sleep that quiets all things
would not take hold of him. He tossed and turned
remembering with pain Patroklos' courage,
his buoyant heart; how in his company
he fought out many a rough day full of danger,
cutting through ranks in war and the bitter sea.
With memory his eyes grew wet. He lay
on his right side, than on his back, and then
face downward -- but at last he rose, to wander
distractedly along the line of surf.
This for eleven nights. The first dawn, brightening
sea and shore, became familiar to him,
as at that hour he yoked his team, with Hektor
tied behind, to drag him out, three times
around Patroklos' tomb. By day he rested
in his own hut, abandoning Hektor's body
to lie full-length in dust -- though Lord Apollo,
pitying the man, even in death,
kept his flesh free of disfigurement.
He wrapped him in his great shield's flap of gold
to save him from laceration. But Achilles
in rage visited indignity on Hektor
day after day, and, looking on,
the blessed gods were moved. Day after day
they urged the Wayfinder steal the body --
a thought agreeable to all but Hera,
Poseidon, and the grey-eyed one, Athena.
These opposed it, and held out, since Ilion
and Priam and his people had incurred
their hatred first, the day Alexandros
made his mad choice and piqued two goddesses,
visitors in his sheepfold: he praised
a third, who offered ruinous lust.
Now when Dawn grew bright for the twelfth day,
Phoibos Apollo spoke among the gods:

"How heartless and how malevolent you are!
Did Hektor never make burnt offering
of bulls' thighbones to you, and unflawed goats?
Even in death you would not stir to save him

for his dear wife to see, and for his mother,
his child, his father, Priam, and his men:
they'd burn the corpse at once and give him burial.
Murderous Achilles has your willing help --
a man who shows no decency, implacable,
barbarous in his ways as a wild lion
whose power and intrepid heart
sway him to raid the flocks of men for meat.
The man has lost all mercy;
he has no shame -- that gift that hinders mortals
but helps them, too. A sane one may endure
an even dearer loss: a blood brother,
a son; and yet, by heaven, having grieved
and passed through mourning, he will let it go.
The Fates have given patient hearts to men.
Not this one: first he took Prince Hektor's life
and now he drags the body, lashed to his car,
around the barrow of his friend, performing
something neither nobler in report
nor better in itself. Let him take care,
or, brave as he is, we gods will turn against him,
seeing him outrage the insensate earth!"

Hera whose arms are white as ivory
grew angry at Apollo. She retorted:

"Lord of the silver bow, your words would be
acceptable if one had a mind to honor
Hektor and Achilles equally.
But Hektor suckled at a woman's breast,
Achilles is the first-born of a goddess --
one I nursed myself. I reared her, gave her
to Peleus, a strong man whom the gods loved,
All of you were present at their wedding --
you too -- friend of the base, forever slippery! --
came with your harp and dined there!"

Zeus the storm king
answered her:

Hera, don't lose your temper
altogether. Clearly the same high honor
cannot be due both men. And yet Lord Hektor,
of all mortal men in Ilion,
was dearest to the gods, or was to me.
He never failed in the right gift; my altar
never lacked a feast

of wine poured out and smoke of sacrifice --
the share assigned as ours. We shall renounce
the theft of Hektor's body; there is no way;
there would be no eluding Achilles' eye,
as night and day his mother comes to him.
Will one of you now call her to my presence?
I have a solemn message to impart:
Achilles is to take fine gifts from Priam,
and in return give back Prince Hektor's body."

And this, Iris who runs on the rainy wind
with word from Zeus departed. Midway between
Samos and rocky Imbros, down she plunged
into the dark grey sea, and the brimming tide
roared over her as she sank into the depth --
as rapidly as a leaden sinker, fixed
on a lure of wild bull's horn, that glimmers down
with a fatal hook among the ravening fish.
Soon Iris came on Thetis in a cave,
surrounded by a company of Nerieds
lolling there, while she bewailed the fate
of her magnificent son, now soon to perish
of Troy's rich earth, far from his fatherland.
Halting before her, Iris said:

          "Come, Thetis,
Zeus of eternal forethought summons you."

Silvery-footed Thetis answered:
          "Why?
Why does the great one call me to him now,
when I am shy of mingling with immortals,
being so heavyhearted? But I'll go.
Whatever he may say will have its weight."

That loveliest of goddesses now put on
a veil so black no garment could be blacker,
and swam where wind swift Iris led. Before them
on either hand the ground swell fell away.
They rose to a beach, then soared into the sky
and found the viewer of the wide world, Zeus,
with all the blissful gods who live forever
around him seated. Athena yielded place,
and Thetis sat down by her father, Zeus,
while Hera handed her a cup of gold
and spoke a comforting word. When she had drunk,
Thetis held out the cup again to Hera.

The father of gods and men began:
        "You've come
to Olympus, Thetis, though your mind is troubled
and insatiable pain preys on your heart.
I know, I too. But let me, even so,
explain why I have called you here. Nine days
of quarreling we've had among the gods
concerning Hektor's body and Achilles.
They wish the Wayfinder make off with it.
I, however, accord Achilles honor
as I now tell you -- in respect for you
whose love I hope to keep hereafter. Go, now,
down to the army, tell this to your son:
the gods are sullen toward him, and I, too,
more than the rest, am angered at his madness,
holding the body by the beaked ships
and not releasing it. In fear of me
let him relent and give back Hektor's body!
At the same time I'll send Iris to Priam,
directing him to go down to the beach head
and ransom his dear son. He must bring gifts
to melt Achilles' rage."

        Thetis obeyed,
leaving Olympus' ridge and flashing down
to her son's hut. She found him groaning there,
inconsolable, while men-at-arms
went to and fro, making their breakfast ready --
having just put to the knife a fleecy sheep.
His gentle mother sat down at his side,
caressed him and said tenderly:

        "My child,
will you forever feed on you own heart
in grief and pain, and take no thought of sleep
or sustenance? It would be comforting
to make love with a woman. No long time
will you live on fore me: Death even now
stands near you, appointed and all-powerful.
But be alert and listen: I am a messenger
from Zeus, who tells me the gods are sullen toward you
and he himself most angered at your madness,
holding the body by the beaked ships
and not releasing it. Give Hektor back.
Take ransom for the body."

Said Achilles:
"Let it be so. Let someone bring the ransom
and take the dead away, if the Olympian
commands this in his wisdom."
      "So, that morning,
in camp, amid the ships, mother and son
conversed together, and their talk was long.
Lord Zeus meanwhile sent Iris to Ilion.

"Off with you, lightfoot, leave Olympus, take
my message to the majesty of Priam
at Ilion. He is to journey down
and ransom his dear son upon the beachhead.
He shall take gifts to melt Achilles' rage,
and let him go alone, no soldier with him,
only some crier, some old man, to drive
his wagon team and guide the nimble wagon,
and afterward to carry home the body
of him that Price Achilles overcame.
Let him not think of death, or suffer dread,
as I'll provide him with a wondrous guide,
the Wayfinder, to bring him across the lines
into the very presence of Achilles.
And he, when he sees Priam within his hut,
will neither take his life nor let another
enemy come near. He is no madman,
no blind brute, nor one to flout the gods,
but dutiful toward men who beg his mercy."

Then Iris at his bidding ran
on the rainy winds to bear the word of Zeus,
until she came to Priam's house and heard
voices in lamentation. In the court
she found the princes huddled around their father,
faces and clothing wet with tears. The old man,
fiercely wrapped and hooded in his mantle,
sat like a figure graven -- caked in filth
his own hands had swept over head and neck
when he lay rolling on the ground. Indoors
his daughters and his sons' wives were weeping,
remembering how many and how brave
the young men were who had gone down to death
before the Argive spearmen.

      Zeus's courier,
appearing now to Priam's eyes alone,
alighted whispering, so the old man trembled:

"Priam, heir of Dardanos, take heart,
and have no fear of me; I bode no evil,
but bring you friendly word from Zeus,
who is distressed for you and pities you
though distant far upon Olympus. He
commands that you shall ransom the Prince Hektor,
taking fine gifts to melt Achilles' rage.
And go alone: no soldier may go with you,
only some crier, some old man, to drive
your wagon team and guide the nimble wagon,
and afterward to carry home the body
of him that Prince Achilles overcame.
Put away thoughts of death, shake off your dead,
for you shall have a wondrous guide,
the Wayfinder, to bring you across the lines
into the very presence of Achilles.
He, for his part, seeing you in his quarters,
will neither take your life nor let another
enemy come near. He is no madman,
no blind brute, nor one to flout the gods,
but dutiful toward men who beg his mercy."

Iris left him, swift as a veering wind.
Then Priam spoke, telling the men to rig
a four-wheeled wagon with a wicker box,
while he withdrew to his chamber roofed in cedar,
high and fragrant, rich in precious things.
He called to Hekabe, his lady:
        "Princess,
word from Olympian Zeus has come to me
to go down to the ships of the Achaians
and ransom our dead son. I am to take
gifts with will melt Achilles' anger. Tell me
how this appears to you, tell me your mind,
for I am torn with longing, now, to pass
inside the great encampment by the ships."

The woman's voice broke as she answered:
        Sorrow,
sorrow. Where is the wisdom now that made you
famous in the old days, near and far?
How can you ever face the Achaian ships
or wish to go alone before those eyes,
the eyes of one who stripped your sons in battle,
how many, and how brave? Iron must be
the heart within you. If he sees you, takes you,
savage and wayward as the man is,

he'll have no mercy and no shame. Better
that we should mourn together in our hall.
Almighty fate spun this thing for our son
the day I bore him: destined him to feed
the wild dogs after death, being far from us
when he went down before the stronger man.
I could devour the vitals of that man,
leeching into his living flesh! He'd know
pain then -- pain like mine for my dead son.
It was no coward the Achaian killed;
he stood and fought for the sweet wives of Troy,
with no more thought of flight or taking cover."

In majesty old Priam said:
        "My heart
is fixed on going. Do not hold me back,
and do not make yourself a raven crying
calamity at home. You will not move me.
If any man on earth had urged this on me --
reader of altar smoke, prophet or priest --
we'd say it was a lie, and hold aloof.
But no: with my own ears I heard the voice,
I saw the god before me. Go I shall,
and no more words. If I must die alongside
the ships of the Achaians in their bronze,
I die gladly. May I but hold my son
and spend my grief; then let Achilles kill me."

Throwing open the lids of treasure boxes
he picked out twelve great robes of state, and twelve
light cloaks for men, and rugs, and equal number,
and just as many capes of snowy linen,
adding a dozen khitons to the lot;
then set in order ten pure bars of gold,
a pair of shining tripods, four great caldrons,
and finally one splendid cup, a gift
Thracians had made him on an embassy.
He would not keep this, either -- as he cared
for nothing now but ransoming his son.

And now, from the colonnade,
he made his Trojan people keep their distance,
berating and abusing them

        "Away,
you craven fools and rubbish! In your own homes
have you no one to mourn, that you crowd here,

to make more trouble for me? Is this a show,
that Zeus has crushed me, that he took the life
of my most noble son? You'll soon know what it means,
as you become child's play for the Achaians
to kill in battle, now that Hektor's gone.
As for myself, before I see my city
taken and ravaged, let me go down blind
to Death's cold kingdom!"

      Staff in hand,
he herded them, until they turned away
and left the furious old man. He lashed out
now at his sons, at Helenos and Paris,
Agathon, Pammon, Antiphonos,
Polites, Deiphobos, Hippothoos,
and Dios -- to these nine the old man cried:

"Bestir yourselves, you misbegotten whelps,
shame of my house! Would god you had been killed
instead of Hektor at the line of ships.
How cursed I am in everything! I fathered
first-rate men, in our great Troy; but now
I swear not one is left: Mestor, Troilos,
laughing amid the war-cars; and then Hektor --
a god to soldiers, and a god among them,
seeming not a man's child, but a god's.
Ares killed them. These poltroons are left,
hollow men, dancers, heroes of the dance,
light-fingered pillagers of lambs and kids
from the town pens!

      Now will you get a wagon
ready for me, and quickly? Load these gifts
aboard it, so that we can take the road."

Dreading the rough edge of their father's tongue,
they lifted out a cart, a cargo wagon,
neat and maneuverable, and newly made,
and fixed upon it a wicker box; then took
a mule yoke from a peg, a yoke of boxwood
 knobbed in front, with rings to hold the reins.
They brought out, too, the band nine forearms long
called the yoke-fastener, and placed the yoke
forward at the shank of the polished pole,
Shoving the yoke pin firmly in. They looped
three turns of the yoke-fastener round the knob.
and wound it over and over down the pole,

tucking the tab end under. Next, the ransom:
bearing the weight of gifts for Hektor's person
out of the inner room, they piled them up
on the polished wagon. It was time to yoke
the mule-team, strong in harness, with hard hooves,
a team the Mysians had given Priam.
Then for the king's own chariot they harnessed
a team of horses of the line of Tros,
reared by the old king in his royal stable.
So the impatient king and his sage crier
had their animals yoked in the palace yard
when Hekabe in her agitation joined them,
carrying in her right hand a golden cup
of honeyed wine, with which, before they left,
they might make offering. At the horses' heads
she stood to tell them:

      "Here, tip wine to Zeus,
the father of gods. Pray for a safe return
from the enemy army, seeing your heart is set
on venturing to the camp against my will.
Pray in the second place to Zeus the storm king,
gloomy over Ida, who looks down
on all Troy country. Beg for an omen-bird,
the courier dearest of all birds to Zeus
and sovereign in power of flight,
that he appear upon our right in heaven.
When you have seen him with your own eyes, then
under that sign, you may approach the ships.
If Zeus who views the wide world will not give you
vision of his bird, then I at least
cannot bid god speed to your journey,
bent on it though you are."
      In majesty
Priam replied:
      "My lady, in this matter
 I am disposed to trust you and agree.
It is an excellent thing and salutary
to lift our hands to Zeus, invoking mercy."

The old king motioned to his housekeeper,
who stood nearby with a basin and a jug,
to pour clear water on his hands. He washed them,
took the cup his lady held, and prayed
while standing there, midway in the walled court.
The he tipped out the wine, looking toward heaven,
saying:

"Zeus, our Father, reigning from Ida,
god of glory and power, grant I come
to Achilles' door as one to be received
with kindliness and mercy. And dispatch
your courier bird, the nearest to your heart
of all birds, and first in power of flight.
Let him appear upon our right in heaven
that I may see him with my own eyes
and under that sign journey to the ships."

Zeus all-foreseeing listened to this prayer
and put an eagle, king
of winged creatures, instantly in flight:
a swamp eagle, a hunter, one they call
the duskwing. Wide as a doorway in a chamber
spacious and high, built for a man of wealth,
a door with long bars fitted well, so wide
spread out each pinion. The great bird appeared
winging through the town on their right hand,
and all their hearts lifted with joy to see him.
In haste the old king boarded his bright car
and clattered out of the echoing colonnade.
Ahead, the mule-team drew the four-wheeled wagon,
driven by Idaios, and behind
the chariot rolled, with horses that the old man
whipped into a fast trot through the town.
Family and friends all followed weeping
as though for Priam's last and deathward ride.
Into the lower town they passed, and reached
the plain of Troy. Here those who followed after
turned back, sons and sons-in-law. And Zeus
who views the wide world saw the car and wagon
brave the plain. He felt a pang for Priam
and quickly said to Hermes, his own son:

Hermes, as you go most happily
of all the gods with mortals, and give heed
to whom you will, be on your way this time
as guide for Priam to the deep sea ships.
Guide him so that not one of the Danaans
may know or see him till he reach Achilles."

Argeiphontes the Wayfinder obeyed.
He bent to tie his beautiful sandals on,
ambrosial, golden, that carry him over water
and over endless land on a puff of wind,
and took the wand with which he charms asleep --

or, when he wills, awake -- the eyes of men.
So, wand in hand, the strong god glittering
paced into the air. Quick as a thought
he came to Helle's waters and to Troy,
appearing as a boy whose lip was downy
in the first bloom of manhood, a young prince,
all graciousness.

        After the travelers
drove past the mound of Ilos, at the ford
they let the mules and horses pause to drink
the running stream. Now darkness had come on
when, looking round, the crier
saw Hermes near at hand. He said to Priam:

"You must think hard and fast, your grace;
there is new danger; we need care and prudence.
I see a man-at-arms there -- ready, I think,
to prey on us. Come, shall we whip the team
and make a run for it? Or take his knees
and beg for mercy?"

        Now the old man's mind
gave way to confusion and to terror.
On his gnarled arms and legs the hair stood up,
and he stared, breathless. but the affable god
came over and took his hand and asked:

        "Old father,
where do you journey, with your cart and car,
while others rest, below the evening star?
Do you not fear the Achaians where they lie
encamped, hard, hostile outlanders, nearby?
Should someone see you, bearing stores like these
by night, how would you deal with enemies?
You are not young, your escort's ancient, too.
Could you beat off an attacker, either of you?
I'll do no hurt to you but defend you here.
You remind me of my father, whom I hold dear."

Old Priam answered him:
        "Indeed, dear boy,
the case is as you say. And yet some god
stretched out his hand above me, he who sent
before me here -- and just at the right time --
a traveler like yourself, well-made, well-spoken,

clearheaded, too. You come of some good family."

The Wayfinder rejoined:
          "You speak with courtesy,
dear sir. But on this point enlighten me:
are you removing treasure here amassed
for safety abroad, until the war is past?
Or can you be abandoning Ilion
in fear, after he perished, that great one
who never shirked a battle, your own princely son?"

Old Priam replied:
          "My brave young friend, who are you?
Born of whom? How nobly you acknowledge
the dreadful end of my unfortunate son."

To this the Wayfinder replied:
          "Dear sir,
you question me about him? Never surmise
I have not seen him with my very eyes,
and often, on the field. I saw him chase
Argives with carnage to their own shipways,
while we stood wondering, forbidden war
by the great anger that Achilles bore
Lord Agamemnon. I am of that company
Achilles led. His own ship carried me
as one of the Myrimidons. My father is old,
as you are, and his name's Polyktor; gold
and other wealth he owns;
and I am seventh and last of all his sons.
When I cast lots among them, my lot fell
to join the siege against Troy citadel.
Tonight I've left the camp to scout this way
where, circling Troy, we'll fight at break of day;
our men are tired of waiting and will not stand
for any postponement by the high command."

Responded royal Priam:
          "If you belong
to the company of Achilles, son of Peleus,
tell me this, and tell me the whole truth:
is my son even now beside the ships?
Or has Achilles by this time dismembered him
and thrown him to the wild dogs?"
          The Wayfinder
made reply again:
          "Dear sir,

no dogs or birds have yet devoured your son.
Beside Achilles' ship, out of the sun,
he lies in a place of shelter. Now twelve days
the man has lain there, yet no part decays,
nor have the blowfly's maggots, that devour
dead men in war, fed on him to this hour.
True that around his dear friend's barrow tomb
Achilles drags him when dawn-shadows come,
driving pitilessly; but he mars him not.
You might yourself be witness, on the spot,
how fresh with dew he lies, washed of his gore,
unstained, for the deep gashes that he bore
have all closed up -- and many thrust their bronze
into his body. The blest immortal ones
favor your prince, and care for every limb
even in death, as they so cherished him."

The old king's heart exulted, and he said:

"Child, it was well to honor the immortals.
He never forgot, at home in Ilion --
ah, did my son exist? was he a dream? --
the gods who own Olympus. They in turn
were mindful of him when he met his end.
Here is a goblet as a gift from me.
Protect me, give me escort, if the gods
attend us, till I reach Achilles' hut."

And in response Hermes the Wayfinder said:
          "You are putting a young man to test
dear sir, but I may not, as you request,
accept a gift behind Achilles' back.
Fearing, honoring him, I could not lack
discretion to that point. The consequence, too,
could be unwelcome. As for escorting you,
even to Argos' famous land I'd ride
a deck with you, or journey at your side.
No cutthroat ever will disdain your guide."

With this, Hermes who lights the way for mortals
leapt into the driver's place. He caught up
reins and whip, and breathed a second wind
into the mule-team and the team of horses.
Onward they ran toward parapet and ships,
and pulled up to the moat.
          Now night had fallen,
bringing the sentries to their supper fire,

but the glimmering god Hermes, the Wayfinder,
showered a mist of slumber on them all.
As quick as thought, he had the gates unbarred
and open to let the wagon enter, bearing
the old king and the ransom.
                Going seaward
they came to the lofty quarters of Achilles,
a lodge the Myrmidons built for their lord
a pine trees cut and trimmed, and shaggy thatch
from mowings in deep meadows. Posts were driven
round the wide courtyard in a palisade,
whose gate one crossbar held, one beam of pine.
It took three men to slam this home, and three
to draw the bolt again -- but great Achilles
worked his entryway alone with ease.
And now Hermes, who lights the way for mortals,
opened for Priam, took him safely in
with all his rich gifts for the son of Peleus.
Then the god dropped the reins, and stepping down
He said:
                "I am no mortal wagoner,
but Hermes, sir. My father sent me here
to be your guide amid the Achaian men.
Now that is done, I'm off to heaven again
and will not visit Achilles. That would be
to compromise an immortal's dignity --
to be received with guests of mortal station.
Go take his knees, and make your supplication:
invoke his father, his mother, and his child;
pray that his heart be touched, that he be reconciled."

Now Hermes turned, departing for Olympus,
and Priam vaulted down. He left Idaios
to hold the teams in check, while he went forward
into the lodge. He found Achilles, dear
to Zeus, there in his chair, with officers
at ease across the room. Only Automedon
and Alkimos were busy near Achilles,
for he had just now made an end of dinner,
eating and drinking, and the laden boards
lay near him still upon the trestles.
                Priam,
the great king of Troy, passed by the others,
knelt down, took in his arms Achilles' knees,
and kissed the hands of wrath that killed his sons.

When, taken with mad Folly in his own land,

a man does murder and in exile finds
refuge in some rich house, then all who see him
stand in awe.
So these men stood.
            Achilles
gazed in wonder at the splendid king,
and his companions marveled too, all silent,
with glances to and fro. Now Priam prayed
to the man before him:
            "Remember your own father,
Achilles, in your godlike youth: his years
like mine are many, and he stands upon
the fearful doorstep of old age. He, too,
is hard pressed, it may be, by those around him,
there being no one able to defend him
from bane of war and ruin. Ah, but he
may nonetheless hear news of you alive,
and so with glad heart hope through all his days
for sight of his dear son, come back from Troy,
while I have deathly fortune.
            Noble sons
I fathered here, but scarce one man is left me.
Fifty I had when the Achaians came,
nineteen out of a single belly, others
born of attendant women. Most are gone.
Raging Ares cut their knees from under them.
And he who stood alone among them all,
their champion, and Troy's, ten days ago
you killed him, fighting for his land, my prince,
Hektor.
            It is for him that I have come
among these ships, to beg him back from you,
and I bring ransom without stint.
            Achilles,
be reverent toward the great gods! And take
pity on me, remember your own father.
Think me more pitiful by far, since I
have brought myself to do what no man else
has done before -- to lift to my lips the hand
of one who killed by son."
            Now in Achilles
the evocation of his father stirred
 new longing, and ache of grief. He lifted
the old man's hand and gently put him by.
Then both were overborne as they remembered:
the old king huddled at Achilles' feet
wept, and wept for Hektor, killer of men,

while great Achilles wept for his own father
as for Patroklos once again; and sobbing
filled the room.
          But when Achilles' heart
had known the luxury of tears, and pain
within his breast and bones had passed away,
he stood then, raised the old king up, in pity
for his grey head and grey beard cheek, and spoke
in a warm rush of words:

          "Ah, sad and old
Trouble and pain you've borne, and bear, aplenty.
Only a great will could have brought you here
 among the Achaian ships, and here alone
before the eyes of one who stripped your sons,
your many sons, in battle. Iron must be
 the heart within you. Come, then, and sit down.
We'll probe our wounds no more but let them rest,
though grief lies heavy on us. Tears heal nothing,
drying so stiff and cold. This is the way
the gods ordained the destiny of men,
to bear such burdens in our lives, while they
feel no affliction. At the door of Zeus
are those two urns of good and evil gifts
that he may choose for us; and one for whom
the lightning's joyous king dips in both urns
will have by turns bad luck and good. But one
to whom he sends all evil -- that man goes
contemptible by the will of Zeus; ravenous
hunger drives him over the wondrous earth,
unresting, without honor from gods or men.
Mixed fortune came to Peleus. Shining gifts
at the gods' hands he had from birth: felicity,
wealth overflowing, rule of the Myrmidons,
a bride immortal at his mortal side.
But then Zeus gave afflictions too -- no family
of powerful sons grew up for him at home,
but one child, of all seasons and of none.
Can I stand by him in his age? Far from my country
I sit at Troy to grieve you and your children.
You, too, sir, in time past were fortunate,
we hear men say. From Makar's isle of Lesbos
northward, and south of Phrygia and the Straits,
no one had wealth like yours, or sons like yours.
Then gods out of the sky sent you this bitterness:
the years of siege, the battles and the losses.
Endure it, then. And do not mourn forever

for your dead son. There is no remedy.
You will not make him stand again. Rather
await some new misfortune to be suffered."
The old king in his majesty replied:

"Never give me a chair, my lord, while Hektor
lies in your camp uncared for. Yield him to me
now. Allow me sight of him. Accept
the many gifts I bring. May they reward you,
and may you see your home again.
You spared my life at once and let me live."

Achilles, the great runner, frowned and eyed him
under his brows:
            "Do not vex me, sir," he said.
"I have intended, in my own good time,
to yield up Hektor to you. She who bore me,
the daughter of the Ancient of the sea,
has come with word to me from Zeus. I know
in your case, too -- though you say nothing, Priam --
that some god guided you to the shipways here.
No strong man in his best days could make entry
into this camp. How could he pass the guard,
or force our gateway?

            Therefore, *let me be*.
Sting my sore heart again, and even here,
under my own roof, suppliant though you are,
I may not spare you, sire, but trample on
the express command of Zeus!"

            When he heard this,
the old man feared him and obeyed with silence.
Now like a lion at one bound Achilles
left the room. Close at his back the officers
Automedon and Alkimos went out --
comrades in arms whom he esteemed the most
after the dead Patroklos. They unharnessed
mules and horses, led the old king's crier
to a low bench and sat him down.
Then from the polished wagon
they took the piled-up price of Hektor's body.
One khiton and two capes they left aside
as dress and shrouding for the homeward journey.
Then, calling to the women slaves, Achilles
ordered the body bathed and rubbed with oil --
but lifted, too, and placed apart, where Priam

could not see his son -- for seeing Hektor
he might in his great pain give way to rage,
and fury then might rise up in Achilles
to slay the old king, flouting Zeus's word.
So after bathing and anointing Hektor
they drew the shirt and beautiful shrouding over him.
Then with his own hands lifting him, Achilles
laid him upon a couch, and with his two
companions aiding, placed him in the wagon.
Now a bitter groan burst from Achilles,
who stood and prayed to his own dead friend:
       "Patroklos,
do not be angry with me, if somehow
even in the world of Death you learn of this --
that I released Prince Hektor to his father.
The gifts he gave were not unworthy. Aye,
and you shall have your share, this time as well."

The prince Achilles turned back to his quarters.
He took again the splendid chair that stood
against the farther wall, then looked at Priam
and made his declaration:
       "As you wished, sir,
the body of your son is now set free.
He lies in state. At the first sight of Dawn
you shall take charge of him yourself and see him.
Now let us think of supper. We are told
that even Niobe in her extremity
took thought for bread -- though all her brood had perished,
her six young girls and six tall sons. Apollo,
making his silver longbow whip and sing,
shot the lads down, and Artemis with raining
arrows killed the daughters -- all this after
Niobe had compared herself with Leto,
the smooth-cheeked goddess.
       She has borne two children,
Niobe said, How many have I borne!
But soon those two destroyed the twelve.
       Besides,
nine days the dead lay stark, no one could bury them,
for Zeus had turned all folk of theirs to stone.
The gods made graves for them on the tenth day,
and then at last, being weak and spent with weeping,
Niobe thought of food. Among the rocks
of Sipylos' lonely mountainside, where nymphs
who race Akheloios river go to rest,
she, too, long turned to stone, somewhere broods on

the gall immortal gods gave her to drink.

Like her we'll think of supper, noble sir.
Weep for your son again when you have borne him
back to Troy; there he'll be mourned indeed."

In one swift movement now Achilles caught
and slaughtered a white lamb. His officers
flayed it, skillful in their butchering
to dress the flesh; they cut bits for the skewers,
roasted, and drew then off, done to a turn.
Automedon dealt loaves into the baskets
on the great board; Achilles served the meat.
Then all their hands went out upon the supper.
When thirst and appetite were turned away,
Priam, the heir of Dardanos, gazed long
in wonder at Achilles' form and scale --
so like the gods in aspect. And Achilles
 in his turn gazed in wonder upon Priam,
royal in visage as in speech. Both men
in contemplation found rest for their eyes,
till the old hero, Priam, broke the silence:

"Make a bed ready for me, son of Thetis,
and let us know the luxury of sleep.
From that hour when my son died at your hands
till now, my eyelids have not closed in slumber
over my eyes, but groaning where I sat
I tasted pain and grief a thousand-fold,
or lay down rolling in my courtyard mire.
Here for the first time I have swallowed bread
and made myself drink wine.
            Before, I could not."
Achilles ordered men and serving women
to make a bed outside, in the covered forecourt,
with purple rugs piled up and sheets outspread
and coverings of fleeces laid on top.
The girls went out with torches in their hands
and soon deftly made up a double bed.
Then Achilles, defiant of Agamemnon,
told his guest:

"Dear venerable sir,
 you'll sleep outside tonight, in case an Achaian
 officer turns up, one of those men
 who are forever taking counsel with me --
            as well they may. If one should see you here

as the dark night runs on, he would report it
to the Lord Marshal Agamemnon. Then
return of the body would only be delayed.
Now tell me this, and give me a straight answer:
How many days do you require
for the funeral of Prince Hektor? -- I should know
how long to wait, and hold the Achaian army."

Old Priam in his majesty replied:

"If you would have me carry out the burial,
Achilles, here is the way to do me grace.
As we are penned in the town, but must bring wood
from the distant hills, the Trojans are afraid.
We should have mourning for nine days in hall,
then on the tenth conduct his funeral
and feast the troops and commons;
on the eleventh we should make his tomb,
and on the twelfth give battle, if we must."

Achilles said:
        "As you command, old Priam,
the thing is done. I shall suspend the war
for those eleven days that you require."

He took the old man's right hand by the wrist
and held it, to ally his fear.
        Now crier
and king with hearts brimful retired to rest
in the sheltered forecourt, while Achilles slept
deep in his palisaded lodge. Beside him,
lovely in her youth, Briseis lay.
And other gods and soldiers all night long,
by slumber quieted, slept on. But slumber
would not come to Hermes the Good Companion,
as he considered how to ease the way
for Priam from the camp, to send him through
unseen by the formidable gatekeepers.
Then Hermes came to Priam's pillow, saying:

"Sir, no thought of danger shakes your rest,
as you sleep on, being great Achilles' guest,
amid men fierce as hunters in a ring.
Your triumphed in a costly ransoming,
but three times costlier your own would be
to your surviving sons -- a monarch's fee --
if this should come to Agamemnon's ear

and all the Achaian host should learn that you are here."

The old king started up in fright, and woke
his herald. Hermes yoked the mules and horses,
took the reins, then inland like the wind
he drove through all the encampment, seen by no one.
When they reached Xanthos, eddying and running
god-begotten river, at the ford,
Hermes departed for Olympus. Dawn
spread out her yellow robe on all the earth,
as they drove on toward Troy, with groans and sighs,
and the mule-team pulled the wagon and the body.
And no one saw them, not a man or woman,
before Cassandra. Tall as the pale-gold
goddess Aphrodite, she had climbed
the citadel of Pergamos at dawn.
Now looking down she saw her father come
in his war-car, and saw the crier there,
and saw Lord Hektor on his bed of death
upon the mule cart. The girl wailed and cried
to all the city:
            "Oh, look down, look down,
go to your windows, men of Troy, and women,
see Lord Hektor now! Remember joy
at seeing him return alive from battle,
exalting all our city and our land!"

Now, at the sight of Hektor, all gave way
to loss and longing, and all crowded down
to meet the escort and body near the gates,
till no one in the town was left at home.
There Hektor's lady and his gentle mother
tore their hair for him, flinging themselves
upon the wagon to embrace his person
while the crowd groaned. All that long day
until the sun went down they might have mourned
in tears before the gateway. But old Priam
spoke to them from his chariot:
            "Make way,
let the mules pass. You'll have your fill of weeping
later, when I've brought the body home."

They parted then, and made way for the wagon,
allowing Priam to reach the famous hall.
They laid the body of Hektor in his bed,
and brought in minstrels, men to lead the dirge.
While these wailed out, the women answered, moaning.

Andromakhe of the ivory-white arms
held in her lap between her hands
the head of Hektor who had killed so many.
Now she lamented:

    "You've been torn from life,
my husband, in young manhood, and you leave me
empty in our hall. The boy's a child
whom you and I, poor souls, conceived; I doubt
he'll come to manhood. Long before, great Troy
will go down plundered, citadel and all,
now that you are lost, who guarded it
and kept it, and preserved its wives and children.
They will be shipped off in the murmuring hulls
one day, and I along with all the rest.
You, my little one, either you come with me
to do some grinding labor, some base toil
for a harsh master, or an Achaian soldier
will grip you by the arm and hurl you down
from a tower here to a miserable death --
out of his anger for a brother, a father,
or even a son that Hektor killed. Achaians
in hundreds mouthed black dust under his blows.
He was no moderate man in war, your father,
and that is why they mourn him through the city.
Hektor, you gave your parents grief and pain
but left me loneliest, and heartbroken.
You could not open your strong arms to me
from your deathbed, or say a thoughtful word,
for me to cherish all my life long
as I weep for you night and day."

    Her voice broke,
and a wail came from the women. Hekabe
lifted her lamenting voice among them:

"Hektor, dearest of sons to me, in life
you had the favor of the immortal gods,
and they have cared for you in death as well.
Achilles captured other sons of mine
in other years, and sold them overseas
to Samos, Imbros, and the smoky island,
Lemnos. That was not his way with you.
After he took your life, cutting you down
with his sharp-bladed spear, he trussed and dragged you
many times round the barrow of his friends,
Patroklos, whom you killed -- though not by this
could that friend live again. But now I find you

fresh as pale dew, seeming newly dead,
like one to whom Apollo of the silver bow
had given easy death with his mild arrows."

Hekabe sobbed again, and the wails redoubled.
Then it was Helen's turn to make lament.

"Dear Hektor, dearest brother to me by far!
My husband is Alexandros,
who brought me here to Troy -- God, that I might
have died sooner! This is the twentieth year
since I left home, and left my fatherland.
But never did I have an evil word
or gesture from you. No -- and when some other
brother-in-law or sister would revile me,
or if my mother-in-law spoke to me bitterly --
but Priam never did, being as mild
as my own father -- you would bring her round
with your kind heart and gentle speech. Therefore
I weep for you and for myself as well,
given this fate, this grief. In all wide Troy
no one is left who will befriend me, none:
they all shudder at me."

        Helen wept,
and a moan came from the people, hearing her.
Then Priam, the old king, commanded them:

"Trojans, bring firewood to the edge or town.
No need to fear an ambush of the Argives.
When he dismissed me from the camp, Achilles
told me clearly they will not harass us,
not until dawn comes for the twelfth day."

Then yoking mules and oxen to their wagons
the people thronged before the city gates.
Nine days they labored, bringing countless loads
of firewood to the town. When Dawn that lights
the world of morals came for the tenth day,
they carried great hearted Hektor out at last,
and all in tears placed his dead body high
upon its pyre, then cast a torch below.
When the young Dawn with finger tips of rose
made heaven bright, the Trojan people massed
about Prince Hektor's ritual fire.
All being gathered and assembled, first
they quenched the smoking pyre with tawny wine

wherever flames had licked their way, then friends
and brothers picked his white bones from the char
in sorrow, while the tears rolled down their cheeks.
In a golden urn they put the bones,
shrouding the urn with veiling of soft purple.
Then in a grave dug deep they placed it
and heaped it with great stones. The men were quick
to raise the death-mound, while in every quarter
lookouts were posted to ensure against
an Achaian surprise attack. When they had finished
raising the barrow, they returned to Ilion,
where all sat down to banquet in his honor
in the hall of Priam king. So they performed
the funeral rites of Hektor, tamer of horses.

Book IX of the Odyssey

Odysseus relates, first, what befell him amongst the Cicones at Ismarus; secondly, amongst the Lotophagi; thirdly, how he was used by the Cyclops Polyphemus.

And Odysseus of many counsels answered him saying: "King Alcinous, most notable of all the people, verily it is a good thing to list to a minstrel such as this one, like to the gods in voice. Nay, as for me, I say that there is no more gracious or perfect delight than when a whole people makes merry, and the men sit orderly at feast in the halls and listen to the singer, and the tables by them are laden with bread and flesh, and a wine-bearer drawing the wine serves it round and pours it into the cups. This seems to me well nigh the fairest thing in the world. But now thy heart was inclined to ask of my grievous troubles, that I may mourn for more exceeding sorrow. What then shall I tell of first, what last, for the gods of heaven have given me woes in plenty? Now, first, will I tell my name, that ye too may know it, and that I, when I have escaped the pitiless day, may yet be your host, though my home is in a far country. I am Odysseus, son of Laertes, who am in men's minds for all manner of wiles, and my fame reaches unto heaven. And I dwell in clear seen Ithaca wherein is a mountain Neriton, with trembling forest leaves, standing manifest to view, and many islands lie around, very near one to the other, Dulichium and Same, and wooded Zacynthus. Now Ithaca lies low, furthest up the sea-line toward the darkness, but those others face the dawning and the sun: a rugged isle, but a good nurse of noble youths; and for myself I can see nought beside sweeter than a man's own country. Verily Calypso, the fair goddess, would fain have kept me with her in her hollow caves, longing to have me for her lord; and likewise too, guileful Circe of Aia, would have stayed me in her halls, longing to have me for her lord. But never did they prevail upon my heart within my breast. So surely is there nought sweeter than a man's own country and his parents, even though he dwell far off in a rich home, in a strange land, away from them that begat him. But come, let me tell thee too of the troubles of my journeying, which Zeus laid on me as I came from Troy.

"The wind that bare me from Ilios brought me nigh to the Cicones, even to Ismarus, whereupon I sacked their city and slew the people. And from out the city we took their wives and much substance, and divided them amongst us, that none through me might go lacking his proper share. Howbeit, thereafter I commanded that we should flee with a swift foot, but my men in their great folly hearkened not. There was much wine still a drinking, and still they slew many flocks of sheep by the seashore and kine with trailing feet and shambling gait. Meanwhile the Cicones went and raised a cry to other Cicones their neighbors, dwelling inland, who were more in number than they and braver withal: skilled they were to fight with men from chariots, and when need was on foot. So they gathered in the early morning as thick as leaves and flowers that spring in their season -- yea and in that hour an evil doom of Zeus stood by us, ill-fated men, that so we might be sore afflicted. They set their battle in array by the swift ships, and the hosts cast at one another with their bronze-shod spears. So long as it was morn and the sacred day waxed stronger, so long we abode their assault and beat them off, albeit they outnumbered us. But when the sun was wending to the time of the loosening of cattle, then at last the Cicones drave in the Achaians and overcame them, and six of my goodly-greaved company perished from each ship: but the remnant of us escaped death and destiny.

"Thence we sailed onward stricken at heart, yet glad as men saved from death, albeit we had lost our dear companions. Nor did my curved ships move onward ere we had called thrice on each of those our hapless fellows, who died at the hands of the Cicones on the plain. Now Zeus, gatherer of the clouds, aroused the North Wind against our ships with a terrible tempest, and covered land and sea alike with clouds, and down sped night from heaven. Thus the ships were driven headlong, and their sails were torn to shreds by the might of the wind. So we lowered the sails into the hold, in fear of death, but rowed the ships landward apace. There for two nights and two days we lay continually, consuming our hearts with weariness and sorrow. But when the fair-tressed Dawn had at last brought the full light of the third day, we set up the masts and hoisted the white sails and sat us down, while the wind and the helmsman guided the ships. And now I should have come to mine own country all unhurt, but the wave and the stream of the sea and the North Wind swept me from my course as I was doubling Malea, and drave me wandering past Cythera.

"Thence for nine whole days was I borne by ruinous winds over the teeming deep; but on the tenth day we set foot on the land of the lotus-eaters, who eat a flowery food. So we stepped ashore and drew water, and straightway my company took their midday meal by the swift ships. Now when we had tasted meat and drink I sent forth certain of my company to go and make search what manner of men they were who here live upon the earth by bread, and I chose out two of my fellows, and sent a third with them as herald. Then straightway they went and mixed with the men of the lotus-eaters, and so it was that the lotus-eaters devised not death for our fellows, but gave them of the lotus to taste. Now whosoever of them did eat the honey-sweet fruit of the lotus, had no more wish to bring tidings nor to come back, but there he chose to abide with the lotus-eating men, ever feeding on the lotus, and forgetful of his homeward way. Therefore I led them back to the ships weeping, and sore against their will, and dragged them beneath the benches, and bound them in the hollow barques. But I commanded the rest of my well-loved company to make speed and go on board the swift ships, lest haply any should eat of the lotus and be forgetful of returning. Right soon they embarked and sat upon the benches, and sitting orderly they smote the grey sea water with their oars.

"Thence we sailed onward stricken at heart. And we came to the land of the Cyclopses, a forward and a lawless fold, who trusting to the deathless gods plant not aught with their hands, neither plough: but, behold, all these things spring for them in plenty, unsown and untilled, wheat, and barley, and vines, which bear great clusters of the juice of the grape, and the rain of Zeus gives them increase. These have neither gatherings for council nor oracles of law, but they dwell in hollow caves on the crest of the high hills, and each one utters the law to his children and his wives, and they reckon not one of another.

"Now there is a waste isle stretching without the harbor of the land of the Cyclopses neither nigh at hand nor yet afar off, a woodland isle, wherein are wild goats unnumbered, for no path of men scares them, nor do hunters resort thither who suffer hardships in the wood, as they range the mountain crests. Moreover, it is possessed neither by flocks nor by ploughed lands, but the soil lies unsown evermore and untilled, desolate of men, and feeds the bleating goats. For the Cyclopses have by them no ships with vermilion cheek, not yet are there shipwrights in the island, who might fashion decked barques, which should accomplish all their desire, voyaging to the towns of men (as oft times men cross the sea to one another in ships), who might likewise have made of their isle a goodly settlement. Yea, it is in no wise a sorry land, but would bear all things in their season; for therein are soft water meadows by the shores of the grey salt sea, and there the vines know no decay, and the land is level to plough; thence might they reap a crop exceeding deep in due season, for verily there is fatness beneath the soil. Also there is a fair haven, where is no need of moorings, either at cast anchor or to fasten hawsers, but men may run the ship on the beach, and tarry until such time as the sailors are minded to be gone, and favorable breezes blow. Now at the head of the harbor is a well of bright water issuing from a cave, and round it are poplars growing. Thither we sailed, and some god guided us through the night, for it was dark and there was no light to see, a mist laying deep about the ships, nor did the moon show her light from heaven, but was shut in with clouds. No man then beheld that island, neither saw we the long waves rolling to the beach, till we had run our decked ships ashore. And when our ships were beached, we took down all their sails, and ourselves too stepped forth upon the strand of the sea, and there we fell into sound sleep and waited for the bright Dawn.

"So soon as early Dawn shone forth, the rosy-fingered, in wonder at the island we roamed over the length thereof: and the Nymphs, the daughters of Zeus, lord of the Aegis, started the wild goats of the hills, that my company might have wherewith to sup. Anon we took to us our curved bows from out the ships and long spear, and arrayed in three bands we began shooting at the goats; and the god soon gave us game in plenty. Now twelve ships bare me company, and to each ship fell nine goats for a portion, but for me alone they set ten apart.

"Thus we sat there the livelong day until the going down of the sun, feasting on abundant flesh and on sweet wine. For the red wine was not yet spent from out the ships, but somewhat was yet therein, for we had each one drawn off large store thereof in jars, when we took the sacred citadel of the Cicones. And we looked across to the land of the Cyclopes who dwell nigh, and to the smoke, and to the voice of the men, and of the sheep and of the goats. And when the sun had sunk and darkness had come on, then we laid us to rest upon the sea beach. So soon as early Dawn shone forth, the rosy-fingered, then I called a gathering of my men, and spake among them all:

"Abide here all the rest of you, my dear companions; but I will go with mine own ship and my ship's company, and make proof of these men, what manner of folk they are, whether froward, and wild, and unjust, or hospitable and of god-fearing mind.'

"So I spake, and I climbed the ship's side, and bade my company themselves to mount, and to loose the hawsers. So they soon embarked and sat upon the benches, and sitting orderly smote the grey sea water with their oars. Now when we had come to the land that lies hard by, we saw a cave on the border near to the sea, lofty and roofed over with laurels, and there many flocks of sheep and goats were used to rest. And about it a high outer court was built with stones, deep bedded, and with tall pines and oaks with their high crown of leaves. And a man was wont to sleep therein, of monstrous size, who shepherded his flocks alone and afar, and was not conversant with others, but dwelt apart in lawlessness of mind. Yea, for he was a monstrous thing and fashioned marvelously, nor was he like to any man that lives by bread, but like a wooded peak of the towering hills, which stands out apart and alone from others.

"Then I commanded the rest of my well-loved company to tarry there by the ship, and to guard the ship, but I chose out twelve men, the best of my company, and sallied forth. Now I had with me a goatskin of the dark wine and sweet, which Maron, son of Euanthes, had given me, the priest of Apollo, the god that watched over Ismarus. And he gave it, for that we had protected him with his wife and child reverently; for he dwelt in a thick grove of Phoebus Apollo. And he made me splendid gifts; he gave me a mixing bowl of pure silver, and furthermore wine which he drew off in twelve jars in all, sweet wine unmingled, as draught divine; nor did any of his servants or of his handmaids in the house know thereof, but himself and his dear wife and one house dame only. And as often as they drank that red wine honey sweet, he would fill one cup and pour it into twenty measures of water, and a marvelous sweet smell went up from the mixing bowl: then truly it was no pleasure to refrain.

"With this wine I filled a great skin, and bare it with me, and corn too I put in a wallet, for my lordly spirit straightway had a boding that man would come to me, a strange man, clothed in mighty strength, one that knew not judgment and justice.

"Soon we came to the cave, but we found him not within; he was shepherding his fat flocks in the pastures. So we went into the cave, and gazed on all that was therein. The baskets were well laden with cheeses, and the folds were thronged with lambs and kids; each kind was penned by itself, the firstlings apart, and the summer lambs apart, apart too the younglings of the flock. Now all the vessels swam with whey, the milk pails and the bowls, the well-wrought vessels where into he milked. My company then spake and besought me first of all to take of the cheeses and to return, and afterwards to make haste and drive off the kids and lambs to the swift ships from out the pens, and to sail over the salt sea water. Howbeit I hearkened not (and far better would it have been), but waited to see the giant himself, and whether he would give me gifts as a stranger's due. Yet was not his coming to be with joy to my company.

"Then we kindled a fire, and made burnt offering, and ourselves likewise took of the cheeses, and did eat, and sat waiting for him within till he came back, shepherding his flocks. And he bore a grievous weight of dry wood, against suppertime. This log he cast down with a din inside the cave, and in fear we fled to the secret place of the rock. As for him, he drove his fat flocks into the wide cavern, even all that he was wont to milk; but the males both of the sheep and of the goats he left without in the deep yard. Thereafter he lifted a huge door stone and weighty, and set it in the mouth of the cave, such an one as two and twenty good four wheeled wains could not raise from the ground, so mighty a sheer rock did he set against the doorway. Then he sat down and milked the ewes and bleating goats all orderly, and beneath each ewe he placed her young. And anon he curdled one half of the white milk, and massed it together, and stored it in wicker baskets, and the other half he let stand in pails, that he might have it to take and drink against supper time. Now when he had done all his work busily, then he kindled the fire anew, and espied us, and made question:

"Stranger, who are ye? Whence sail ye over the wet ways? On some trading enterprise or at adventure do ye rove, even as sea robbers over the brine, for at hazard of their own lives they wander, bringing bale to alien men."

"So spake he, but as for us our heart within us was broken for terror of the deep voice and his own monstrous shape; yet despite all I answered and spake unto him, saying:

"Lo, we are Achaeans, driven wandering from Troy, by all manner of winds over the great gulf of the sea; seeking our homes we fare, but another path have we come, by other ways: even such, methinks, was the will and the counsel of Zeus. And we avow us to be the men of Agamemnon, son of Atreus, whose fame is even now the mightiest under heaven so great a city did he sack, and destroyed many people; but as for us we have lighted here, and come to these thy knees, if perchance thou wilt give us a stranger's gift, or make any present, as is the due of strangers. Nay, lord, have regard to the gods, for we are thy suppliants: and Zeus is the avenger of suppliants and sojourners, Zeus, the god of the stranger, who fareth in the company of reverend strangers.'

"So I spake, and anon he answered out of his pitiless heart: 'Thou art witless, my stranger, or thou hast come from afar, who bids me either to fear or shun the gods. For the Cyclopes pay no heed to Zeus, lord of the Aegis, nor to the blessed gods, for verily we are better men than they. Nor would I, to shun the enmity of Zeus, spare either thee or they company, unless my spirit bade me. But tell me where thou didst stay thy well-wrought ship on thy coming? Was it perchance at the far end of the island, or hard by, that I may know?'

"So he spake tempting me, but he cheated me not, who knew full much, and I answered him again with words of guile:

"'As for my ship, Poseidon, the shaker of the earth, brake it to pieces, for he cast it upon the rocks at the border of your country, and brought it nigh the headland, and a wind bare it thither from the sea. But I with these my men escaped from utter doom.'

"So I spake, and out of his pitiless heart he answered me not a word, but sprang up, and laid his hands upon my fellows, and clutching two together dashed them as they had been whelps, to the earth, and the brain flowed forth upon the ground, and the earth was wet. Then cut he them up piecemeal, and made ready his supper. So he ate even as a mountain-bred lion, and ceased not, devouring entrails and flesh and bones with their marrow. And we wept and raised our hand to Zeus, beholding the cruel deeds; and we were at our wits' end. And after the Cyclops had filled his huge maw with human flesh and the milk he drank thereafter, he lay within the cave, stretched out among his sheep.

"So I took counsel in my great heart, whether I should draw near, and pluck my sharp sword from my thigh, and stab him in the breast, where the midriff holds the liver, feeling for the place with my hand. But my second thought withheld me, for so should we too have perished even there with utter doom. For we should not have prevailed to roll away with our hands from the lofty door the heavy stone which he set there. So for that time we made moan, awaiting the bright Dawn.

"Now when early Dawn shone forth, the rosy-fingered, again he kindled the fire and milked his goodly flocks all orderly, and beneath each ewe set her lamb. Anon when he had done all his work busily, again he seized yet another two men and made ready his mid-day meal. And after the meal, lightly he moved away the great door stone, and drove his fat flocks forth from the cave, and afterwards he set it in his place again, as one might set the lid on a quiver. Then with a loud whoop, the Cyclops turned his fat flocks toward the hills; but I was left devising evil in the deep of my heart, if in any wise I might avenge me, and Athene grant me renown.

"And this was the counsel that showed best in my sight. There lay by a sheepfold a great club of the Cyclops, a club of olive wood, yet green, which he had cut to carry with him when it should be seasoned. Now when we saw it we likened it in size to the mast of a black ship of twenty oars, a wide merchant vessel that traverses that great sea gulf, so huge it was to view in bulk and length. I stood thereby and cut off from it a portion as it were a fathom's length, and set it by my fellows, and bade them file it down, and they made it even, while I stood by and sharpened it to a point, and straightway I took it and hardened it in the bright fire. Then I laid it well away, and hid it beneath the dung, which was scattered in great heaps in the depths of the cave. And I bade my company cast lots among them, which of them should risk the adventure with me, and lift the bar and turn it about in his eye, when sweet sleep came upon him. And the lot fell upon those four whom I myself would have been fain to choose, and I appointed myself to be the fifth among them. In the evening he came shepherding his flocks of goodly fleece, and presently he drove his fat flocks into the cave each and all, nor left he any without in the deep court yard, whether through some foreboding, or perchance that the god so bade him do. Thereafter he lifted the huge door stone and set it in the mouth of the cave, and sitting down he milked the ewes and bleating goats, all orderly, and beneath each ewe he place her young. Now when he had done all his work busily, again he seized yet other two and made ready his supper. Then I stood by the Cyclops and spake to him, holding in my hands an ivy bowl of the dark wine:

"'Cyclops, take and drink wine after thy feast of man's meat, that thou may know what manner of drink this was that our ship held. And lo, I was bringing it thee as a drink offering, if haply thou may take pity and send me on my way home, but thy mad rage is past all sufferance. O hard of heart, how may another of the many men there be come ever to thee again, seeing that thy deeds have been lawless?'

"So I spake, and he took the cup and drank it off, and found great delight in drinking the sweet draught, and asked me for it yet a second time:

"'Give it me again of thy grace, and tell me thy name straightway, that I may give thee a stranger's gift, wherein thou may be glad. Yea for the earth, the grain giver, bears for the Cyclops the mighty clusters of the juice of the grape, and the rain of Zeus gives them increase, but this is a rill of very nectar and ambrosia.'

"So he spake, and again I handed him the dark wine. Thrice I bare and gave it him, and thrice in his folly he drank it to the lees. Now when the wine had got about the wits of the Cyclops, then did I speak to him with soft words:

"'Cyclops, thou ask me my renowned name, and I will declare it unto thee, and do thou grant me a stranger's gift, as thou didst promise. Noman is my name, and Noman they call me, my father and my mother and all my fellows.'

"So I spake, and straightway he answered me out of his pitiless heart.

"'Noman will I eat last in the number of his fellows, and the others before him: that shall be they gift.'

"Therewith he sank backwards and fell with face upturned, and there he lay with his great neck bent round, and sleep, that conquers all men, overcame him. And wine and the fragments of men's flesh issued forth from his mouth, and he vomited, being heavy with wine. Then I thrust in that stake under the deep ashes, until it should grow hot, and I spake to my companions comfortable words, lest any should hang back from me in fear. But when that bar of olive wood was just about to catch fire in the flame, green though it was, and began to glow terribly, even then I came nigh, and drew it from the coals, and my fellows gathered about me, and some god breathed great courage into us. For their part they seized the bar of olive wood, that was sharpened at the point, and thrust it into his eye, while I from my place aloft turned it about, as when a man bores a ship's beam with a drill while his fellows below spin it with a strap, which they hold at either end, and the auger runs round continually. Even so did we seize the fiery pointed brand and whirled it round in his eye, and the blood flowed about the heated bar. And the breath of the flame singed his eyelids and brows all about, as the ball of the eye burnt away, and the roots thereof crackled in the flame. And as when a smith dips an axe or adze in chill water with a great hissing, when he would temper it - for hereby anon comes the strength of iron -- even so did his eye hiss round the stake of olive. And he raised a great and terrible cry, that the rock rang around, and we fled away in fear, while he plucked forth from his eye the brand bedabbled in much blood. Then maddened with pain he cast it from him with his hands, and called with a loud voice to the Cyclops, who dwelt about him in the caves along the windy heights. And they heard the cry and flocked together from every side, and gathering round the cave and asked him what ailed him:

"'What hath so distressed thee, Polyphemus, that thou criest thus aloud through the immortal night, and make us sleepless? Surely no mortal drives off thy flocks against thy will: surely none slays thyself by force of craft?

"And the strong Polyphemus spake to them again from out the cave: 'My friends, Noman is slaying me by guile, not at all by force.'

"And they answered and spake winged words: 'If then no man is violently handling thee in thy solitude, it can in no wise be that thou should escape the sickness sent by mighty Zeus. Nay, pray thou to thy father, the lord Poseidon.'

"On this wise they spake and departed; and my heart within me laughed to see how my name and cunning counsel had beguiled them. But the Cyclops, groaning and travailing in pain, groped with his hands, and lifted away the stone from the door of the cave, and himself sat in the entry, with arms outstretched to catch, if he might, any one that was going forth with his sheep, so witless, methinks, did he hope to find me. But I advised me how all might be for the very best, if perchance I might find a way of escape from death for my companions and myself, and I wove all manner of craft and counsel, as a man will for his life, seeing that great mischief was nigh. And this was the counsel that showed best in my sight. The rams of the flock were well nurtured and thick of fleece, great and goodly, with wool dark as the violet. Quietly I lashed them together with twisted withies, whereon the Cyclops slept, that lawless monster. Three together I took: now the middle one of the three would bear each a man, but the twain went on either side, saving my fellows, Thus every three sheep bare their man. But as for me I laid hold of the back of a young ram who was for the best and the goodliest of all the flock, and curled beneath his shaggy belly there I lay, and so clung face upward, grasping the wondrous fleece with a steadfast heart. So for that time making moan we awaited the bright Dawn.

"So soon as early Dawn shone forth, the rosy-fingered, then did the rams of the flock hasten forth to pasture, but the ewes bleated un-milked about the pens, for their udders were swollen to bursting. Then their lord, sore stricken with pain, felt along the backs of all the sheep as they stood up before him, and guessed not in his folly how that my men were bound beneath the breasts of his thick-fleeced flocks. Last of all the sheep came forth the ram, cumbered with his wool, and the weight of me and my cunning. And the strong Polyphemus laid his hands on him and spake to him, saying:

"'Dear ram, wherefore, I pray thee, art thou the last of all the flocks to go forth from the cave, who of old was not wont to lag behind the sheep, but were ever the foremost to pluck the tender blossom of the pasture, faring with long strides, and were still the first to come to the streams of the rivers, and first did long to return to the homestead in the evening. But now art thou the very last. Surely though art sorrowing for the eye of thy lord, which an evil man blinded, with his accursed fellows, when he had subdued my wits with wine, even Noman, whom I say hath not yet escaped destruction. Ah, if thou couldst feel as I, and be endued with speech, to tell me where he shifts about to shun my wrath; then should he be smitten, and his brains be dashed against the floor here and there about the cave, and my heart be lightened of the sorrows which Noman, nothing worth, hath brought me!'

"Therewith he sent the ram forth from him, and when we had gone but a little way from the cave and from the yard, first I loosed myself from under the ram and then I set my fellows free. And swiftly we drove on those stiff-shanked sheep, so rich in fat, and often turned to look about, till we came to the ship. And a glad sight to our fellows were we that had fled from death, but the others they would have bemoaned with tears; howbeit I suffered it not, but with frowning brows forbade each man to weep. Rather I bade them to cast on board the many sheep with goodly fleece, and to sail over the salt sea water. So they embarked forthwith, and sat upon the benches, and sitting orderly smote the grey sea water with their oars. But when I had not gone so far, but that man's shout might be heard, then I spoke unto the Cyclops taunting him:

"'Cyclops, so thou were not to eat the company of a weakling by main might in thy hollow cave! Thine evil deeds were very sure to find thee out, thou cruel man, who had no shame to eat thy guests within thy gates, wherefore Zeus hath requited thee, and the other gods.'

"So I spake, and he was mightily angered at heart, and he brake off the peak of a great hill and threw it at us, and it fell in front of the dark-prowed ship. And the sea heaved beneath the fall of the rock, and the backward flow of the wave bare the ship quickly to the dry land, with the wash from the deep sea, and drove it to the shore. Then I caught up a long pole in my hands, and thrust the ship from off the land, and roused my company, and with a motion of the head bade them dash in with their oars, that so we might escape our evil plight. So they bent to their oars and rowed on. But when we had now made twice the distance over the brine, I would fain have spoken to the Cyclops, but my company stayed me on every side with soft words, saying:

"Foolhardy that thou art, why would you rouse a wild man to wrath, who even now has cast so mightily a throw towards the deep and brought our ship back to land, yea, and we thought that we had perished even there? If he had heard any of us utter sound or speech he would have crushed our heads and our ship timbers with a cast of a rugged stone, so mightily he hurls.'

"So spake they, but they prevailed not on my lordly spirit, and I answered him again from out of an angry heart:

"'Cyclops, if any one of mortal men shall ask thee of the unsightly blinding of thine eye, say that it was Odysseus that blinded it, the waster of cities, son of Laertes, whose dwelling is in Ithaca.

"So I spake, and with a moan he answered me, saying:

"'Lo now, in very truth the ancient oracles have come upon me. There lived here a soothsayer, a noble man and a mighty, Telemus, son of Eurymus, who surpassed all men in soothsaying, and waxed old as a seer among the Cyclops. He told me that all these things should come to pass in the aftertime, even that I should lose my eyesight at the hand of Odysseus. But I ever looked for some tall and goodly man to come hither, clad in great might, but behold now one that is a dwarf, a man of no worth and a weakling, hath blinded me of my eye after subduing me with wine. Nay come hither, Odysseus, that I may set by thee a stranger's cheer, and speed they parting hence, that so the Earth-shaker may vouchsafe it thee, for his son am I , and he avows him for my father. And he himself will heal me, if it be his will; and none other of the blessed gods or of mortal men.'

"Even so he spake, but I answered him, and said:

"Would god that I were as sure to rob thee of soul and life, and send thee within the house of Hades, as I am that not even the Earth-shaker will heal thine eye!

"So I spake, and then he prayed to the lord Poseidon stretching forth his hands to the starry heaven: "Hear me, Poseidon, girdler of the earth, god of the dark hair, if indeed I be thine, and thou avowest thee my sire, -- grant that he may never come to his home, even Odysseus, waster of cities, the son of Laertes, whose dwelling is in Ithaca: yet if he is ordained to see his friends and come unto his well-built house, and his own country, late may he come in evil case, with the loss of all his company, in the ship of strangers, and find sorrows in his house.'

"So he spake in prayer, and the god of the dark locks heard him. And once again he lifted a stone, far greater than the first, and with one swing he hurled it, and he put forth a measureless strength, and cast it but a little space behind the dark-prowed ship, and all but struck the end of the rudder. And the sea heaved beneath the fall of the rock, but the wave bare on the ship and drove it to the further shore.

"But when we had now reached that island, where all our other decked ships abode together, and our company were gathered sorrowing, expecting us evermore, on our coming thither we ran our ship ashore upon the sand, and ourselves too stepped forth upon the sea beach. Next we took forth the sheep of the Cyclops from out the hollow ship, and divided them, that none through me might go lacking his proper share. But the ram for me alone my goodly-greaved company chose out, in the dividing of the sheep, and on the shore I offered him up to Zeus, even to the son of Cronos, who dwells in the dark clouds, and is lord of all, and I burnt the slices of the thighs. But he heeded not the sacrifice, but was devising how my decked ships and my dear company might perish utterly. Thus for that time we sat the livelong day, until the going down of the sun, feasting on abundant flesh and sweet wine. And when the sun had sunk and darkness had come on, then we laid us to rest upon the sea beach. So soon as early Dawn shone forth, the rosy-fingered, I called to my company, and commanded them that they should themselves climb the ship and loose the hawsers. So they soon embarked and sat upon the benches, and sitting orderly smote the grey sea water with their oars.

"Thence we sailed onward stricken at heart, yet glad as men saved from death, albeit we had lost our dear companions.

From the *Odyssey* by Homer, written somewhere between 1050 and 850 BC
Translated by Butcher and Lang

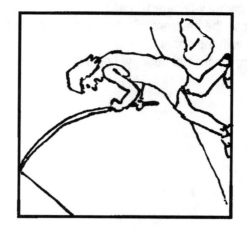

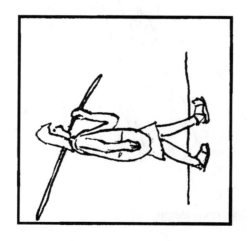

L-18a: Scythian Origin Myth

A golden bowl is the centerpiece of a Scythian origin myth, as reported by the Greeks. The bowl, brimming with wine and blood, fell from the sky before three nomads. Flames erupted each time one of the nomads tried to drink from the bowl. Only the youngest of them could tolerate the fire and drink the sacred mixture. He became the first Scythian king and united the nomads into the Great Kingdom. The young king proclaimed the gold bowl to be the sacred vessel of life, and, filled with blood and wine, it was central to most Scythian rituals. The bowl symbolized the power of a patron goddess; the wine represented wealth and abundance; and the blood meant death to the enemies of the Scythians.

L-18b: The Amazons and the Scyths

It is reported of the Sauromatae, that when the Greeks fought with the Amazons, whom the Scythians called Oior-pata or "man-slayers," as it may be rendered, Oior being Scythic for "man," and pata for "to slay"- it is reported, I say, that the Greeks, after gaining the battle of the Thermodon, put to sea, taking with them on board three of their vessels all the Amazons whom they had made prisoners; and that these women upon the voyage rose up against the crews, and massacred them to a man. As, however, they were quite strange to ships, and did not know how to use rudder, sails, or oars, they were carried, after the death of the men, where the winds and the waves listed. At last they reached the shores of the Palus Maeotis and came to a place called Cremni or "the Cliffs," which is in the country of the free Scythians. Here they went ashore, and proceeded by land towards the inhabited regions; the first herd of horses which they fell in with they seized, and mounting upon their backs, fell to plundering the Scythian territory.

The Scyths could not tell what to make of the attack upon them. The dress, the language, the nation itself, were unknown. Whence the enemy had come, even, was a marvel. Imagining, however, that they were all men of about the same age, they went out against them, and fought a battle. Some of the bodies of the slain fell into their hands, whereby they discovered the truth. Hereupon they deliberated, and made a resolve to kill no more of them, but to send against them a detachment of their youngest men, as near as they could guess equal to the women in number, with orders to encamp in their neighborhood, and do as they saw them do - when the Amazons advanced against them, they were to retire, and avoid a fight- when they halted, the young men were to approach and pitch their camp near the camp of the enemy. All this they did on account of their strong desire to obtain children from so notable a race. So the youths departed, and obeyed the orders that had been given them. The Amazons soon found out that they had not come to do them any harm; and so they on their part ceased to offer the Scythians any molestation. And now, day after day, the camps approached nearer to one another; both parties led the same life, neither having anything but their arms and horses, so that they were forced to support themselves by hunting and pillage.

At last an incident brought two of them together - the man easily gained the good graces of the woman, who bade him by signs (for they did not understand each other's language) to bring a friend the next day to the spot where they had met- promising on her part to bring with her another woman. He did so, and the woman kept her word. When the rest of the youths heard what had taken place, they also sought and gained the favor of the other Amazons.

The two camps were then joined in one, the Scythians living with the Amazons as their wives; and the men were unable to learn the tongue of the women, but the women soon caught up the tongue of the men. When they could thus understand one another, the Scyths addressed the Amazons in these words- "We have parents, and properties, let us therefore give up this mode of life, and return to our nation, and live with them. You shall be our wives there no less than here, and we promise you to have no others." But the Amazons said - "We could not live with your women - our customs are quite different from theirs. To draw the bow, to hurl the javelin, to bestride the horse, these are our arts; of womanly employments we know nothing. Your women, on the contrary, do none of these things; but stay at home in their wagons, engaged in womanish tasks, and never go out to hunt, or to do anything. We should never agree together. But if you truly wish to keep us as your wives, and would conduct yourselves with strict justice towards us, go home to your parents, bid them give you your inheritance, and then come back to us, and let us and you live together by ourselves."

The youths approved of the advice, and followed it. They went and got the portion of goods which fell to them, returned with it, and rejoined their wives, who then addressed them in these words following: "We are ashamed, and afraid to live in the country where we now are. Not only have we stolen you from your fathers, but we have done great damage to Scythia by our ravages. As you like us for wives, grant the request we make of you. Let us leave this country together, and go and dwell beyond the Tanais." Again the youths complied.

Crossing the Tanais they journeyed eastward a distance of three days' march from that stream, and again northward a distance of three days' march from the Palus Maeotis. Here they came to the country where they now live, and took up their abode in it. The women of the Sauromatae have continued from that day to the present to observe their ancient customs, frequently hunting on horseback with their husbands, sometimes even unaccompanied; in war taking the field; and wearing the very same dress as the men.

The Sauromatae speak the language of Scythia, but have never spoke it correctly, because the Amazons learned it imperfectly at the first. Their marriage law lays it down that no girl shall wed till she has killed a man in battle. Sometimes it happens that a woman dies unmarried at an advanced age, having never been able in her whole lifetime to fulfill the condition.

From Book IV, The History of Herodotus

⌐ ⌐ ⌐ ⌐ ⌐ ⌐ ⌐ ⌐ ⌐ ⌐ ⌐ ⌐ ⌐ ⌐ ⌐

Literature Helps

This section includes helpful definitions of literary terms.

STORY ELEMENTS

<u>Plot</u>: The plot is the plan of action in a story, the actions that tell the reader what is happening to the characters and why. This series of events is called the plot line.

<u>Plot or story line</u> shows the actions or series of events in the story. It has five parts: exposition, rising action, climax, falling action, and resolution.

<u>Episodic Plot</u>: An episodic plot is several short adventures or episodes within one long story.

<u>Exposition</u> is that part of the play or story, usually the beginning, that explains the background and setting of the story and often introduces the characters.

<u>Rising Action</u> is the central part of a story during which problems arise, leading up to the climax.

<u>Climax</u>: The climax is the highest point of action and interest in a story. It is also the turning point where the story begins to resolve.

<u>Falling Action</u> is that part of the story that follows the climax or turning point; it contains the action or dialogue necessary to lead the story to a resolution or ending.

<u>Resolution</u>: The resolution poses the solution to the problem and begins to tie up the story.

<u>Conclusion</u>: The conclusion sums up the message or theme of the story bringing it to its end.

⌐ ⌐ ⌐ ⌐ ⌐ ⌐ ⌐ ⌐ ⌐ ⌐ ⌐ ⌐ ⌐ ⌐

<u>Theme</u>: The theme is the message that the author is communicating, the topic.

<u>Mood</u>: The mood is the "feeling" of a story created by powerful words.

<u>Setting</u> is the time and place that the action of a story takes place.

<u>Moral</u> is the lesson the author is trying to teach in the story.

<u>Conflict</u> is the problem in a story that has to be solved. A story usually shows one of two kinds of conflict:

> <u>Internal</u>: This shows the main character having a struggle within himself.

> <u>External</u>: This shows the main character having a struggle with someone or something other than himself – man against nature, man against man, man against society, and man against the unknown.

A story may have more than one type of conflict.

<u>Character</u>: Characters may either be flat or round.

      Flat – a flat character is one-dimensional, they never change

      Round – this character is well-developed, has depth, and goes through many changes

<u>Protagonist</u> is the hero of the story

<u>Antagonist</u> is the person or thing that fights against the hero.

<u>Motivation</u> is a reason that explains a character's thoughts, feelings, actions, or speech.

POINT OF VIEW

First person: A story that is told by one character and uses the word "I".

Third person: A story told by one character using pronouns such as Him or Them.

Three types of third person stories.

1. <u>Omniscient</u>: (unlimited) This tells the thoughts, ideas, and feelings of any or all the characters.

2. <u>Limited Omniscient</u>: This sees the action through the eyes of usually just one character.

3. <u>Objective</u>: This does not enter into the mind of the character. The story is told through the actions not the thoughts of the characters.

FIGURES OF SPEECH

A Figure of Speech is a figurative expression that is imaginative rather than factual. It compares things that are not alike in reality but are alike in the author's imagination. Some common expressions are as follows.

Personification: This is figurative language that gives human characteristics to an object.

      Ex. The leaves of the trees clap their hands.

Simile: compares one thing to a different thing using *like* or *as*.

      Ex. He sang as a bird in spring.

Metaphor: compares one thing with a different thing without using *like* or *as*.

      Ex. You're a barrel of laughs.

Irony: a literary expression in which the intended meaning of the words are opposite their usual sense.

      Ex. I just love to lose a race.

To state something as fact that turns out not to be true is another example of irony. Irony can also occur in a story when something is obvious to the reader but not to the characters. Another form of irony is when the author uses words that are too weak (understatement), that play down the situation, or too strong (overstatement), that exaggerate the situation. Irony also takes place when the characters say or do something that they think is normal, but you, the reader, know is not quite the way it should be. Sarcasm is another form of irony that says exactly the opposite of what is meant. A good reader will be able to recognize that the author's words and intentions are not necessarily the same.

Hyperbole: exaggeration for effect, not to be taken literally
>Ex. He could have reached the moon with that throw.

Idiom: an expression that has a special meaning which is different from the usual meaning of the individual words.
>Ex. David drew a blank on the last question of his science quiz.

WRITING TECHNIQUES

Symbolism: takes on a meaning in the work far beyond its ordinary meaning. It is anything that stands for an object, idea, or an emotion other than itself. Examples of Symbolism include allegory, fable, and parable. Symbolism is the use of extended metaphors.

Imagery: Imagery gives the reader a clear mental picture of what is happening, often in the form of a vivid description or a striking comparison. It may be a metaphor, simile, or a straightforward description. The use of Imagery may also have a symbolic meaning.

Foreshadowing: to shadow or typify beforehand, urging the reader on to future events in the story

Flashback: an interruption in the progression of a story by telling an earlier event

Malapropism: a funny misuse of one word for a similar one
>Ex. I just ate a gigantic hamburglar for lunch.

Satire: a literary work in which vices, follies, etc., are held up to ridicule and scorn

Allusion: a reference to a well-known person, place, event, literary work, or work of art

Repetition: repeating key words or phrases to build suspense, excitement, fear, or for emphasis

Dialogue: written conversations between the characters in the story.

RHETORIC
Rhetoric defined by Plato -- "A universal art of winning the mind by arguments, which means not merely arguments in the courts of justice, and all other sorts of public councils, but in probate conference as well."

1. Rhetoric prevents the triumph of fraud and injustice. It is not enough to know the truth but to be able to argue for the right decision before others.

2. Rhetoric is a method of instruction for the public. A speaker has to "educate" the audience by framing arguments with the help of common knowledge and commonly accepted opinions.

3. Rhetoric makes us see both sides of a case. You are most prepared to refute an opponent's arguments by understanding the position they come from.

4. Rhetoric is a means of defense. Aristotle said, "If it is a disgrace to a man when he cannot defend himself in a bodily way, it would be odd not to think him disgraced when he cannot defend himself with reason. Reason is more distinctive than is bodily effort." Often those coming from a position of reason can avoid violent confrontation.

LEVELS OF ARGUMENT SOUNDNESS

<u>Fact</u>: something that is known to be true

<u>Opinion</u>: something that is thought to be true; someone's personal views about the facts

<u>Allegation</u>: a stated assertion, especially without proof

<u>Biased Information</u>: information that is told from only one point of view

<u>Propaganda</u>: allegations, facts, opinions, biased information and the like presented with the intention of making you believe a certain way

Examples include:
    <u>Name Calling</u>: the use of labels to arouse hatred or disgust for a person, group, or idea

    <u>Glittering Generalities</u>: the use of phrases such as "wonderful", "the best", "the worlds most...."

    <u>Testimonial</u>: having someone who allegedly has experience with a product or idea speak on its behalf

    <u>Bandwagon</u>: the use of phrases such as "everybody's doing it" to convince you that you need to do it, too

    <u>Superstition</u>: using people's fears to persuade them to act or think in a certain way

TYPES OF LITERATURE

Literature is classified as fiction or non-fiction.

Fiction: Fiction includes imaginative Prose and Poetry. Prose fiction can be further subdivided into Prose Narrative and Drama.

<u>Prose Narrative</u> includes Novel, Novelette, and Short Story that are further divided into five categories: realistic, biographical, historical, fanciful, and science fiction.
Fanciful fiction includes myths, legends, tall tales, fables, and fairytales.

<u>Drama</u> can be classified as Prose or Poetic; both can be in the form of Comedy or Tragedy.
Comedies are most commonly classified as Classical, Tragi-comedy, Melodrama, Romance, Farce, and Fantasy.

Tragedies are classified as Classical, Historical, or Problem Play.

Poetry may be either Narrative or Lyric.

<u>Narrative Poetry</u> includes Epic, Ballad, and Metrical Tale

    Epic – celebrates the feats of a legendary or national hero

    Ballad – often of folk origin, consisting of simple stanzas and usually having a recurrent refrain

Metrical Tale – a tale composed in verse; sung or chanted in meter

Free Verse - poetry not written in a regular pattern or meter.

<u>Lyric Poetry</u> includes General, Dramatic, Pastoral, Sonnet, Ode, and Elegy

Dramatic – a monologue revealing a character, addressed directly to the reader

Pastoral – portrays rural life usually in an idealized manner

Sonnet – consists of 14 lines; 8 lines embody the statement, 4 the resolution of a single theme

Ode – lengthy, often addressed to a praised object, person, or quality

Elegy – expresses sorrow; mournful, melancholy, pensive in tone

Haiku - three line Japanese verse form; the first and third lines have five syllables. The second line has seven syllables.

Limerick - humorous, rhyming, five-line poem with a specific meter and rhyme

Blank Verse - poetry written in unrhymed iambic pentameter lines.

Non-fiction: Non-fiction, or non-imaginative prose, is based on fact and true situations. It may, however, be creative and entertaining. Non-fiction includes biography, autobiography, articles, textbooks, brochures, booklets, letters, essays, and newspapers. Non-fiction prose is divided into four categories or modes:

<u>Exposition</u> – The writer presents information, explaining something to his reader. His intent is to explain, analyze, present information, or clarify a subject. Expository writing is used in directions, instructions, informative reports, interviews, letters, summaries, news stories, biographies, analysis or interpretation.

<u>Description</u> – The writer evokes a place, object, or character. Description is used to present the physical details, or an impression or an object or an individual.

<u>Narration</u> – The writer tells a story about himself or others. Narration is used to recount details of an event. Narration is used in short stories, summaries, and news stories.

<u>Persuasion or Argument</u> – The writer tries to convince people that they should agree with his position on some issue. Persuasion is used to arouse in the reader a need to act or arrive at a given conclusion.
This type of writing appeals to the person's emotions and is used for editorials, movie reviews, book reports, letters to senators, etc.

These modes rarely appear alone in non-fiction prose. Modes tend to shift within a work, telling stories to enhance arguments, describing something before explaining its function, etc. Exposition is the most common mode of non-fiction prose. The essay is a form of non-fiction prose with a purpose to explain or inform. Essays can be grouped

6

into three categories: those that instruct, those that recount, and those that analyze. They can be written in either formal or informal style.

WRITING STYLE
Formal: This sounds old-fashioned; it is usually precise and descriptive.

Informal: This is more modern; it often sounds as if someone were talking to you.

Informational: This will be concise; any description it gives is factual.

DEFINITIONS
Legend – a story passed down by oral tradition popularly believed to have historical basis

Myth – a traditional tale serving to explain some natural phenomenon

Proverb – a short saying expressing an obvious truth

Parable – a short simple story that teaches a moral lesson

Fable – usually about animals and teaches a moral lesson

Allegory - narrative in which abstract ideas are personified; a description to convey a different meaning from that which is expressed

Anecdote - brief story about an interesting, amusing, or strange event

Fairy tale – a fantasy, an unbelievable or untrue story

Folk tale - story composed orally and passed from person to person by word of mouth

Melodrama - drama written in an exaggerated way to produce strong feeling or excitement in the reader

Farce - literature written specifically to make the reader laugh

Tribute – in literature, something said to show gratitude, respect, honor, or praise

Oration – a formal speech usually given at a ceremony

Rhetoric – the art of using words effectively; especially the art of prose composition

Style – manner of expression characteristic of the author or to the work

Technique – the tools an author uses to create a work

Analogy – comparing things, such as two pairs of words. The word pairs should be related in the same manner. Analogies can be stated two ways:

> Hammer is to nail as screwdriver is to screw.
> Hammer:nail::screwdriver:screw
> One colon stands for "is to" and two colons stand for "as".

BASIC ELEMENTS OF ESSAY WRITING

The purpose of the essay is to explain or inform. The essayist is a person with a purpose; an idea he wants to communicate. The essay is a method of logical thought presentation. It should be organized in a clear understandable progression: introduction, body, and conclusion. Keep in mind your essay deals with one subject but can vary in length. Essays can be grouped into four categories: those that describe, those that persuade, those that recount, and those that analyze or interpret. They can be written in either formal or informal style.

Your opening paragraph, the introduction, should state your topic and how you are going to approach it. Each ensuing paragraph should detail individual points brought up in your introduction. The final paragraph should conclude or close your thoughts or feelings on the topic.

GRADING THE ESSAY

Emphasis should primarily be on content when grading the essay and should account for at least fifty percent of the grade. The essay should meet the guidelines described above. It should deal with one topic and follow a logical thought pattern. The opening paragraph should state clearly the topic and the body of the essay should speak only to the topic. The conclusion should restate the topic summing up with thoughts or feelings on it.

Emphasis should next be placed on sentence structure; this should account for about twenty-five percent of the grade. *Elements in Style* and *Elements in Grammar*, edited by E.B. White, are helpful tools for building writing skills. Watch for incomplete sentences, run on sentences, and sentence fragments.

Grammar and spelling together should account for twenty-five percent of the grade.

When using *The Institute for Excellence in Writing* follow the grading guide on page 26.